Concrete structures

Concrete structures

Protection, repair and rehabilitation

R. Dodge Woodson

AMSTERDAM • BOSTON • HEIDELBERG • LONDON
NEW YORK • OXFORD • PARIS • SAN DIEGO
SAN FRANCISCO • SINGAPORE • SYDNEY • TOKYO
Butterworth-Heinemann is an imprint of Elsevier

Butterworth-Heinemann is an imprint of Elsevier
30 Corporate Drive, Suite 400, Burlington, MA 01803, USA
Linacre House, Jordan Hill, Oxford OX2 8DP, UK

Library of Congress Cataloging-in-Publication Data
Application submitted

British Library Cataloguing-in-Publication Data
A catalogue record for this book is available from the British Library.

ISBN: 978-1-85617-549-4

For information on all Butterworth-Heinemann publications
visit our Web site at www.elsevierdirect.com

Typeset by Macmillan Publishing Solutions
www.macmillansolutions.com

Printed in the United States of America
09 10 11 12 13 10 9 8 7 6 5 4 3 2 1

This book is dedicated to Adam, Afton, and Leona for sticking with me as I worked on this project. They have been patient and supportive throughout the process.

Contents

Acknowledgments

I would like to thank the following sources for providing art and illustrations contained in this book:

The United States Government
Reed Pumps
Faddis Concrete Products

About the author

R. Dodge Woodson has worked in the field of construction for over 30 years. During most of those years he has owned his own business. Woodson has written dozens of professional reference books for people in the trades, served as an expert witness in construction cases, and enjoyed consulting with new business owners. His name is known internationally as an expert in the field of construction.

R. Dodge Woodson lives on substantial acreage in Maine and enjoys outdoor activities. Some of his favorite pastimes are photography, fly fishing, target shooting, four wheeling, and metal detecting.

Introduction

If you work with concrete, then you need this detailed reference guide. Repairing and rehabilitating concrete is big business, but it is a specialized type of work. Rehab work is not the same as new construction. There are many considerations for protecting existing concrete from damage and bonding new concrete to old concrete properly. You are holding an indispensible tool when it comes to working with the combination of new and old concrete: this book.

R. Dodge Woodson is an internationally known expert in the field of construction. He has written dozens of professional reference books and worked for over 30 years in the field as a contractor and tradesman. Additionally, attorneys call upon Woodson to serve as an expert witness in cases involving construction matters. Without a doubt, Woodson is the foremost authority on the subject of construction.

You are probably reading this book for a particular reason. Are you looking for answers to your concrete questions? You will find them here. Are you wondering about the concrete code? The quizzes in this book will help you check your knowledge. So look over the table of contents, thumb through the pages, and take a look at the many illustrations. Pay particular attention to the tip boxes that bring key information to your fingertips quickly.

What does a single mistake in the field cost you? If you are an employee, it could cost you your job. Self-employed contractors pay for their mistakes and the mistakes made by their employees. Woodson can tell you from experience that mistakes are costly and can be extremely expensive. If this book keeps you from making one mistake during your entire career, it paid for itself. Realistically, you should not go back into the field without this definitive guide to working with concrete. It won't teach you everything you will ever need to know about concrete, but it will reveal sides of the business that you will not find anyplace else. Consider this book an insurance policy. You have already chosen your career path, and this book is the map that will make your journey more productive.

Introduction

CHAPTER 1

Understanding concrete

Time takes its toll on concrete structure, which creates a problem for the use of concrete in a country's infrastructure. Knowing the right principles and procedures for the repair and rehabilitation of concrete structures is a critical element to financial success. Tearing down existing structures and rebuilding them from the ground up can be cost prohibitive. Learning and perfecting the ways to make the most of existing infrastructures are key elements when it comes to sustainable living and safe living conditions.

Many people look at concrete and see nothing but, well, concrete. But the knowledgeable mind sees much more. Are there stress cracks in the surface? Were expansion joints installed properly? Does the color of the concrete indicate a proper curing time? Is the surface a slick, glasslike finish or a brushed finish? Is the material flaking away? Can existing flaws be repaired in such a way to guarantee structural integrity?

Most people take concrete for granted. Yet, it is one of the strongest building blocks of many bridges, highways, and other significant infrastructure. Working with a new installation of concrete is very different from repairing and rehabilitating existing concrete structures. Both types of work have their rules of thumb and their engineering elements. It often requires more experience to repair concrete than it does to install it as new construction. This is what you will learn here.

Are you capable of doing a site inspection and evaluation of existing concrete structures? If not, would you like to be? This is your chance to learn a great deal about existing concrete structures. A few experts may be needed for complicated compromises in concrete construction, but many situations can be assessed personally.

How much do you know about your options when it comes to fixing problems with concrete? Would you like to know more? Of course you would! And this book tells you what to use, when to use it, and how to use it.

What is the best way to avoid the need for concrete repairs? Maintenance is a great place to start, and this procedure will be discussed in detail. If you are looking for specifics on concrete work and codes, you have come to the right place.

doi: 10.1016/B978-1-85617-549-4.00001-0

So what are you going to learn? Well, here are just a few topics:

Determining the quality of concrete
High-performance concrete
In-place strength of concrete
Pervious concrete used for sustainable building projects
Recycling concrete
Durability testing
Performance specifications and data
Lithium technology
The effects of fly ash
Admixtures
Certifications
Quality control
Control charting
LEED-accredited concrete procedures
Using recycled water in ready-mixed concrete
Predictive testing
Measured compressive strength
Aggregates
Green concrete
Self-sealing concrete
Thermal characteristics
High-belite cement dam concrete
Crack resistance
Mechanical behavior
Secular distortion performance
Corrosion inhibitors
Posttension concrete
Prestressed concrete
Reinforcing steel in concrete structures

Okay, now you know what this book covers, so let's do it.

CHAPTER 2

Evaluating concrete in concrete structures

Proper evaluation of concrete requires several steps. The requirements for repairs or rehabilitation are generally established in one of three ways. In the worst case, there is a failure of structural integrity that prompts the need for an evaluation. Visual inspections often reveal a need for further evaluation and testing. If cracks, flaking, or other visual defects are encountered, a full investigation and evaluation are usually warranted. The third way to establish the need for an evaluation is periodic testing. This is the safest way to check the strength and dependability of concrete structures.

REVIEWING THE RECORDS

Construction documents are typically filed before the permits for construction are issued. In the case of concrete construction, reviewing original documents and a paper trail over the history of the structure is one step in a thorough evaluation. Here are some examples of existing records that should be available for consideration:

- Design documents
- Plans
- Specifications
- History reports
- Inspection reports
- Site inspection results
- Laboratory test records
- Concrete records on the materials used in construction and the batch plant
- Instrumentation documents
- Operation reports
- Maintenance reports
- Monument survey data

doi: 10.1016/B978-1-85617-549-4.00002-2

SITE SURVEY

A site survey consists of a visual exam of all exposed concrete. The inspectors are looking for any implication of distress that may require repair or rehabilitation. The standard procedure is to create a map of potential defects that are encountered. These could include cracks, disintegration, spalling, or joint deterioration. The mapping process is typically done with foldout sketches of the monolith surfaces. Mapping must include inspection and delineating of pipe and electrical galleries, filling, and emptying, culverts when possible. Core drilling is a common procedure when evaluating concrete. The material recovered from the drilling can be taken to a laboratory for scientific testing.

It is not acceptable to use tests for splitting tensile strength to establish the acceptance of concrete in the field.

SHODDY WORKMANSHIP

Shoddy workmanship can lead to significant problems in concrete structures. Site inspections can bring such problems to light and result in further testing. Some of the normal defects that are looked for in a visual inspection include the following:

- Cold joints
- Bug holes
- Reinforcing steel that has become exposed
- Honeycombing
- Surface defects that can indicate serious problems

Workmanship should be monitored during the construction process. In a perfect world, it would be, but in the real world, corners are cut, and problems can result from the lack of professionalism. Since safety is a paramount concern, routine inspections are needed to confirm that the workmanship in a structure is as it should be.

Sometimes different materials are used on the same job for different portions of work. When this is the case, each material must be evaluated for the purpose it is being used for.

CRACKING

Cracking is a common problem in concrete construction. Homeowners see it in basement floors, garage floors, and basement walls. Cracks occur in sidewalks, dams, bridges, and retaining walls. Any crack is a reason for concern and warrants a thorough inspection and investigation.

Table 2.1 Terms Associated with Visual Inspection of Concrete

Construction faults
  Bug holes
  Cold joints
  Exposed reinforcing steel
  Honeycombing
  Irregular surface
	Cracking
	  Checking or crazing
	  D-cracking
	  Diagonal
	  Hairline
	  Longitudinal
	  Map or pattern
	  Random
	  Transverse
	  Vertical
	  Horizontal

Disintegration
  Blistering
  Chalking
  Delamination
  Dusting
  Peeling
  Scaling
  Weathering
	Distortion or movement
	  Buckling
	  Curling or warping
	  Faulting
	  Settling
	  Tilting

Erosion
  Abrasion
  Cavitation	

Joint-sealant failure	

Seepage
  Corrosion
  Discoloration or staining
  Exudation
  Efflorescence
  Incrustation
	Spalling
	  Popouts
	  Spall

Courtesy of United States Army Corps of Engineers

Pattern cracking

Pattern cracks are common. These cracks tend to be short and uniformly distributed throughout a concrete surface. Pattern cracking can have two causes: It can indicate restraint of contraction on the surface layer by the backing or inner concrete, or it can be due to an increase in the volume in the interior concrete.

You may hear pattern cracking referred to as map cracks, crazing, checking, or D-cracking. D-cracking is often found in the lower part of a concrete slab, usually near a joint in the concrete. If you find moisture accumulation, you could find D-cracking.

Isolation cracks

Isolation cracks appear as individual cracks. This type of cracking indicates tension on the concrete. The tension is usually perpendicular to the cracks. An individual crack can run in a diagonal, longitudinal, transverse, vertical, or horizontal direction.

> Code enforcement officers have the authority to require the results of strength tests of cylinders cured under field conditions.

Crack depth

Crack depth is categorized into four terms: surface, shallow, deep, and through.

Crack width

Crack width ranges from fine to medium to wide. Fine cracks are typically less than 0.04 inch wide. A medium crack would be between 0.04 to 0.08 inch. Wide cracks exceed 0.08 inch.

> When preparing to pour concrete, installers must do the following:
>
> - Clean all equipment.
> - Make sure no debris is in the concrete.
> - Make sure no ice is in the concrete.
> - Clean the forms.
> - Make sure any filler units that are in contact with the concrete are well drenched.
> - Make sure there are no deleterious coatings or ice on the reinforcing materials.
> - Make sure there is no water in the path of the concrete installation without the consideration and approval of a code officer.
> - Make sure no unsound material is present.

Crack activity

Crack activity has to do with the presence of a particular factor causing a crack. Determining crack activity is necessary to determine a mode or repair. If the cause of a crack is causing more cracking, then the crack is active. Any crack that is currently moving is considered active. If a specific cause for a crack cannot be determined, the crack must be considered active.

Dormant cracks do not have current movement. Some cracks are considered dormant when any movement of the crack is minimal enough to not interfere with a repair plan.

> When concrete is mixed, it must be mixed to a uniform distribution of materials.

ALL CRACK OCCURRENCE

Cracks can occur before or after concrete cures. The cracking can be structural or nonstructural. It may be hard to determine whether a crack is structural or nonstructural by visual inspection only. A full analysis by a structural engineer is normally required to make a full determination of the type of cracking that is encountered.

Structural cracks tend to be wide. Their openings can increase as a result of continuous loading and creep of the concrete. As a rule of thumb, any crack that could be structural in nature should be treated as a structural defect and receive a full evaluation from appropriate experts.

Table 2.2 Vibration Limits for Freshly Placed Concrete (Hulshizer and Desci 1984)

Age of Concrete at Time of Vibration (hr)	Peak Particle Velocity of Ground Vibrations
Up to 3	102 mm/sec (4.0 in./sec)
3 to 11	38 mm/sec (1.5 in./sec)
11 to 24	51 mm/sec (2.0 in./sec)
24 to 48	102 mm/sec (4.0 in./sec)
Over 48	178 mm/sec (7.0 in./sec)

Courtesy of United States Army Corps of Engineers

DISINTEGRATION AND SPALLING

Disintegration is the deterioration by any means of concrete. The mass of particles being removed from the main body of concrete distinguishes disintegration from spalling. When the loss is small, you are dealing with disintegration. A large loss of intact concrete is considered spalling.

When concrete it poured, the final deposit method must not separate or lose materials in the process.

SCALING

Scaling is a form of disintegration. A common cause of scaling is freezing and thawing conditions. Localized flaking or peeling is normally a form of scaling. Scaling, which can also be referred to as spalling, is rated based on the depth of the defect. The rating system is as follows:

- *Light spalling* is when the loss of surface mortar does not expose any coarse aggregate.

- *Medium spalling* occurs when the damage has a depth of up to 0.4 inch.
- *Severe spalling* is determined when the depth of the damage ranges from 0.2 to 0.4 inch. There is also some loss of mortar surrounding aggregate particles with a depth range of 0.4 to 0.8 inch.
- *Very severe spalling* is the loss of coarse aggregate particles, as well as surface mortar and surrounding aggregate, generally to a depth greater than 0.8 inch.

DUSTING

Dusting occurs when there is a development of a powdered material at the surface of hardened concrete. Horizontal concrete surfaces that receive heavy traffic are the most likely surfaces to find dusting occurring. Poor workmanship practices are usually the cause of the problem. Anyone who has worked in construction for any length of time has seen finishers spray water on a concrete surface during the finishing process. This often causes dusting.

Concrete that is being poured should be installed as closely as practical to the final resting place of the mixture.

DISTORTION

Distortion of concrete is also known as movement. Put simply, this is a change in alignment of the components of a structure. This could be wall movement or support movement. When evaluating distortion, historic data is likely to be helpful. Assuming that good records have been maintained over the years, your research may turn up a history of continuing failure due to distortion.

EROSION

Erosion is a common concern in concrete construction. Experts generally divide erosion into two categories: abrasion and cavitation. Abrasion is recognized by the smooth surface it leaves behind on concrete. This is due to repeated rubbing and grinding of debris, equipment, gravel, or other items against concrete. Repeated impact forces that are caused by a collapse of a vapor bubble in rapidly flowing water cause cavitation. Erosion caused by water does not generally leave a smooth surface on concrete. A rough, pitted concrete surface is a sign of cavitation. In severe cases, cavitation can result in structural damage. Statistics indicate that cavitation generally requires a water velocity of at least 40 feet per second.

Vertically formed lifts must be level.

JOINT SEALS AND SEEPAGE

Joint seals can fail. The seals are intended to repel water. If a seal fails and water invades a concrete joint, buckling, cracking, erosion, and other problems may occur. Another purpose of joint seals is to prevent debris from entering an expansion joint. When debris embeds in a joint, it can result in a failure with the expansion joint.

Seepage consists of water or other fluids moving through pores or interstices. When conducting a visual inspection, an inspector can check for seepage by looking for the following:

- Water
- Dampness
- Moisture
- Corrosion
- Discoloration
- Staining
- Exudations
- Efflorescence
- Incrustations

Seepage is common around hydraulic structures. Any seepage found should be reported. These data are necessary to maintain historical facts for future review. Cracks that are associated with hardening concrete fall into two categories: those that occur while the concrete is hardening and those that occur after the concrete has hardened. Knowing when a crack occurs and what caused the cracking is always of interest to those who evaluate concrete construction.

Concrete must remain plastic and flow readily when it is being installed.

SPECIAL CASES OF SPALLING

Two special cases involve spalling. The first type is known as a popout. These defects are shallow and are usually conical depressions in the surface of the concrete. When concrete is poured in freezing conditions, especially if some unsatisfactory aggregate particles are present, you may find popouts.

A popout is created when water finds its way into coarse aggregate and then freezing begins. The ice pushes off the top of the aggregate particle and the superjacent layer of mortar. This leaves shallow pitting. Certain materials are more susceptible to popouts than others. If you are working with chert particles of low specific gravity, limestone that contains clay, or shaly material, you may see spalling.

The second case of spalling involves the corrosion of reinforcement material. A visual inspection for this type of defect is fairly simple. Evidence will be exposed reinforcement materials that are protruding through the concrete. Rust staining on the reinforcement material is also present in many cases.

It is common practice to maintain a minimum temperature for most concrete installations at 50°F.

DELAMINATION

Delamination can occur when reinforcing steel is installed too close to the surface of a concrete structure. Chloride ions or air that create rust or iron oxide will corrode steel reinforcement material. A corresponding increase in volume up to eight times the original volume amount is possible. The increased volume can crack concrete.

A simple, inexpensive method is available to test for delamination. All you need are a pair of safety glasses and a hammer. Tap the concrete with the hammer to check for defects. If you hear a sharp "ping," it is a good sign that delamination is not present. However, a hollow, echo sound means more tests are needed. This type of test is common when working with small surface areas.

Larger surface areas require extensive time to check with a hammer. Horizontal surfaces can be tested by dragging a chain over the area. Listen for that same "ping" sound. This is a simple way to test more concrete in less time.

Infrared thermography is a more advanced method for inspecting concrete for delamination. The thermal gradients within concrete that is exposed to sunlight can be measured with thermography equipment. Delamination interrupts the heat transfer through concrete. Higher surface temperatures will be present if delamination exists. Infrared thermography is capable of identifying and recording areas that are affected by delamination.

Accelerated curing is an allowable practice.

CRACK SURVEYS

Crack surveys are needed for historical records and to expose potential problems. Obviously, cracks are not supposed to exist in concrete structures. When they are found, the cracks must be identified. Typically, cracks are marked and researched. Once the type and cause of cracks are known, they can be assessed and recorded for future review.

SIZING CRACKS

Cracks can be sized with several methods. A simple card that contains lines of various widths can be used to estimate crack size. Crack monitors, which are small, hand-held microscopes, are used for more specific measurements. There are also transducers that can be used for crack measuring. Once the size of a crack is determined, it should be recorded in the historical data of the concrete structure. By doing this, the cracks can be monitored and measured to determine if they are growing in size.

Obtaining the depth of cracks is often simple, but in some cases, it can be extremely difficult. A small measuring device, such as a feeler gauge, can be used to establish the depth of some cracks. It is not uncommon for simple methods to fail in determining the depth of a crack. When this is the case, inspectors must turn to more sophisticated methods, such as drilling or pulse-velocity measurements.

> At no time should forms, fillers, or the ground with which the concrete will come into contact be covered with frost.

SURFACE MAPPING

Surface mapping is important when establishing the history of a concrete surface over time. The types of defects that are sought in surface mapping may include the following:

- Cracking
- Spalling
- Scaling
- Popouts
- Honeycombing
- Exudation
- Distortion
- Unusual discoloration
- Erosion
- Cavitation
- Seepage
- Joint condition
- Joint materials
- Corrosion of reinforcement materials

Inspectors who are going to perform surface mapping by hand will need certain tools:

- Structural drawings
- Historical data
- Documentation tools, such as a notebook computer or a notepad and pen

- Tape measure
- Ruler
- Feeler gauge
- Hand microscope
- Knife
- Hammer
- Fine wire
- String
- Flashlight
- Camera outfit
- Tape recorder

JOINT INSPECTIONS

Joint inspections can be done with a visual tour of all joints. Expansion, contraction, and construction joints should all be inspected. The condition of joints, both good and bad, should be noted for inclusion in the historical record. Some potential defects to look for include spalling, D-cracking, chemical attacks, seepage, and any emission of solids.

CORE DRILLING

Core drilling is the best method for testing concrete, but it is very expensive. If the quality of concrete in a structure is suspected to be weakened with general inspections, core drilling may be called for. Scaling, leaching, or pattern cracking can be signs of the need for core drilling.

How deep does core drilling go? It depends on the structure. For example, a massive structure may require core sampling to be done at a depth of up to 2 feet. The diameter of a core sample should be at least three times the nominal maximum size of aggregate. When there is little mortar bonding the concrete across the diameter of the core, you are likely to wind up with rubble rather than a solid sample. Core samples must be properly labeled, oriented, and stored for future observation. Written records are also required to maintain consistency in the historical data.

There are times when a bore hole camera is helpful. The use of such an instrument can reveal facts about the inner condition of a concrete structure. For example, if your core samples are coming up as rubble, the bore hole camera may be your best alternative.

> All forms used for concrete installations are required to prevent the leakage of mortar.

UNDERWATER INSPECTIONS

Underwater inspections are usually conducted by scuba divers. When a very deep or long dive is required, a diver with surface-supplied air is a better option. Flexibility and speed is an advantage for the scuba diver. In clear water, a visual inspection can be done. Many types of structures are located in water that is not clear enough to perform a visual examination.

Fortunately, many types of test devices that are used above water have been adopted for use below water. Rebound hammers work underwater. Both direct and indirect ultrasonic pulse-velocity systems can be used below the water surface. These tools can give a diver a good reading on the general condition of concrete that is surrounded by water.

Underwater vehicles

Underwater vehicles are commonly used to inspect submerged concrete structures. These vehicles come in five different categories of manned units: untethered, tethered, diver lockout, observation or work bells, and atmospheric diving suits. All are operated by a person inside, who has viewing ports, dry conditions, and some degree of mobility.

Unmanned vehicles are another option for underwater inspections. These include tethered, free swimming, towed, towed midwater, bottom-reliant, bottom-crawling, structurally reliant, and ntethered. Unmanned vehicles are known as remotely operated vehicles (ROVs). Television cameras are mounted on the vehicles. Control of the vehicle is done from the water surface with a navigation system, such as a joystick. These vehicles can be fitted up to perform inspections and maintenance.

ROVs offer the advantage of being operated at extreme depths. They can remain underwater for long periods of time. Repeated tasks can be completed accurately with ROVs. Another advantage is that ROVs can be operated in harsh conditions that would hamper general diving operations.

> All concrete forms must be installed and secured in a manner that ensures the forms will maintain the desired position and shape.

When ROVs are compared to manned vehicles, there are pros and cons. For example, manned vehicles are big, bulky, and expensive to operate. ROVs are small, flexible, and relatively inexpensive. An ROV provides a two-dimensional view, while a manned unit can provide three-dimensional assessments. Both types of vehicles have their place in underwater inspections.

Photographic tools

Photographic tools have come a long way over the years. An underwater inspection can involve the use of either still cameras or video cameras, or both. Video systems can see through turbid water conditions. This is a big plus over the eyes of a diver.

HIGH-RESOLUTION ACOUSTIC MAPPING SYSTEM

High-resolution acoustic mapping systems can be used to check for erosion and faulting. These systems consist of three basic components: the positioning subsystem, the acoustic subsystem, and the compute-and-record subsystem. An acoustic subsystem is made up of a boat-mounted transducer array and signal-processing electronics. This type of system sends output back to a computer. The computer calculates the elevation of the bottom surface from the information supplied.

A lateral positioning subsystem has a sonic transmitter on a boat and two or more transponders in the water at a known or surveyed location. The transponders receive a sonic pulse from the transmitter. This information is radioed to the survey vessel. A time and location is determined by the survey vessel.

Compute-and-record subsystems provide computer-controlled operation of the system and for processing, display, and storage of data. Real-time mapping is done in a computerized manner.

While high-resolution systems are extremely accurate, they do have limitations. These systems typically work in depths ranging from 5 to 40 feet. Another drawback is that a high-resolution system works best in calm water. If there is wave activity that exceeds 5 degrees, a hi-res system will shut down.

When removing concrete forms, workers must be sure to maintain all safety and serviceability of the concrete structure.

SIDE SCANNER

A side scanner sonar requires two transducers mounted in a waterproof housing. When a signal is sent from the scanner, it is called a sonograph. Darkened areas and shadows are used for evaluation. The width of shadows and the position of objects can be used to calculate height. Newer versions of scanners have far fewer limitations than the earlier models did. Side scanners have been proven useful in breakwaters, jetties, groins, port structures, and inland waterway facilities, such as locks and dams.

OTHER MEANS OF UNDERWATER TESTING

Other means of underwater testing include radar, ultrasonic pulse velocity, ultrasonic pulse-echo systems, and sonic pulse-echo techniques for piles. All of these methods have their advantages. Let's look at some of them.

Advantages of radar systems

- The electromagnetic signal emitted from radar travels very quickly.
- Conductivity controls the loss of energy and, therefore, the penetration depth.
- Dielectric constant determines the propagation velocity.

Advantages of ultrasonic pulse velocity

- Provides a nondestructive method for evaluating structures.
- Measures the time of travel of acoustic pulses of energy through a material of known thickness.
- Piezoelectric transducers are housed in metal casings and are excited by high-impulse voltages as they transmit and receive acoustic pulses.
- An oscilloscope in the system measures time and displays acoustic waves.
- Reliable in situ delineations of the extent and severity of cracks, areas of deterioration, and general assessments.
- Capable of penetrating up to 300 feet of continuous concrete with the aid of amplifiers.
- Can be transported easily.
- Has a high data acquisition to cost ratio.
- Can be converted for underwater use.

Advantages of ultrasonic pulse-echo systems

- Uses piezoelectric crystals to generate and detect signals and the accurate time base of an oscilloscope to measure the time of arrival of a longitudinal ultrasonic pulse in concrete.
- These systems can delineate sound concrete, concrete of questionable quality, deteriorated concrete, delaminations, voids, reinforcing steel, and other objects within concrete.
- Can determine the thickness of concrete up to about 1.5 feet.
- Can be adapted to water environments.

Advantages of pulse-echo techniques for piles

- Can determine the length of concrete piles, in tens of feet, in dry soil or underwater.
- Uses a round-trip echo time in the pile to measure an accurate time base of an oscilloscope.
- Can be used to calculate the reference between length and diameter ratios.

LABORATORY WORK

A lot can be done with site visits and visual inspections. However, it is often laboratory work that tells the tale of the tape. Petrographic exams use a branch of geology that deals with the descriptions and classifications of rocks. Hardened concrete is considered a synthetic sedimentary rock. Petrographic exams check for the following:

- Aggregate condition
- Pronounced cement-aggregate reactions
- Deterioration of aggregate particles in place
- Denseness of cement paste
- Homogeneity of concrete
- Settlement and bleeding of fresh concrete
- Depth and extent of carbonation
- Occurrence and distribution of fractures
- Characteristics and distribution of voids
- Presence of contaminating substances

CHEMICAL ANALYSIS

Chemical analysis of hardened concrete can be used to estimate the cement content, original water–cement ratio, and the presence and amount of chloride and other admixtures. This is another form of testing that is part of the larger puzzle in determining the qualities of concrete.

PHYSICAL ANALYSIS

Physical analysis is often done on core samples. This testing looks for nine elements:

- Density
- Compressive strength
- Modulus of elasticity
- Poisson's ratio
- Pulse velocity
- Direct shear strength of concrete bonded to foundation rock
- Friction sliding of concrete on foundation rock
- Resistance of concrete to deterioration caused by freezing and thawing
- Air content and parameters of the air-void system

NONDESTRUCTIVE TESTING

Nondestructive testing (NDT) is used to determine various relative properties of concrete. Strength, modulus of elasticity, homogeneity, and integrity of concrete can be calculated with NDT. There are many approaches to NDT, and they require inspectors to have expertise in the given approach to arrive at accurate data.

Rebound hammers

Earlier, we talked a little about rebound hammers. This is a form of NDT and a fast and simple way to test concrete. However, the test is imprecise and cannot accurately predict the strength of concrete. Some factors that can skew a test with a rebound hammer include the following:

- Smoothness of a concrete surface
- Moisture content
- Type of course aggregate
- Size, shape, and rigidity of specimen

Carbonation of concrete surface probes

Probes can be used to do NDT. The probe may use a powder cartridge to insert a high-strength steel probe into a section of concrete. The results of probe measurements can be converted to compressive strength values. There are reports, however, that probes can sometimes supply inaccurate data.

A probe is normally used to test density. A probe will embed deeper in concrete that is suffering from failure in density, subsurface hardness, and as the strength of concrete weakens. This type of testing is fine for on-site, general tests, but it is limited. Precise measurements are not available from probe testing. The act of probing concrete will leave a hole in the concrete surface that must be repaired.

Ultrasonic pulse-velocity testing

Ultrasonic pulse-velocity testing is probably the most frequently used means of NDT. The results of this testing can be calculated. High velocities indicate good concrete, while low velocities reveal weak concrete. The system for this testing is portable and can penetrate about 35 linear feet of concrete. Testing of this type is fast. However, an inspector must have access to opposite sides of the section being tested, and this can present a problem.

Acoustic mapping

Acoustic mapping provides comprehensive evaluation of the top surface wear of concrete in such structures as aprons, sills, lock chamber floors, and so forth. Fast, accurate evaluations of horizontal sections below water can be done with acoustic mapping. Dewatering is not needed. Accuracy falls off at depths greater than 30 feet.

Ultrasonic pulse-echo testing

Ultrasonic pulse-echo testing is good for flat surfaces. It can detect steel and plastic pipe that is embedded in concrete. Resolution is good with this type of testing equipment. Improvements in this form of testing continue to develop.

Radar

Radar is an NDT. It does not require contact with concrete. Resolution and penetration is somewhat limited. Some opinions favor signal testing over radar, but radar is a growing element in concrete evaluation.

OTHER CONSIDERATIONS

Some other considerations when evaluating concrete include stability analysis, deformation monitoring, concrete service life, and reliability analysis.

Well, there you have it! Now that we have discussed the evaluation process, let's move on to Chapter 3.

CHAPTER

Causes of distress and deterioration of concrete

3

The list of potential causes of distress and deterioration of concrete is a long one. A few examples include chemical reactions, shrinkage, weathering, and erosion. Many other potential causes exist, and we will explore them individually. Understanding the factors that can damage concrete structures is an important element in the business of rehabilitation and repair work.

ACCIDENTAL LOADINGS

Accidental loadings are not common. (This is why they are accidental.) When an earthquake occurs and affects concrete structures, that action is considered an accidental loading. This type of damage is generally short in duration and few and far between in occurrences.

A visual inspection is likely to reveal spalling or cracking when accidental loadings occur. Unfortunately, this type of damage cannot be prevented because the causes are not expected and are difficult to prepare for. For example, an engineer does not expect a ship to hit a piling on a bridge, but it happens. The only defense is to build with as much caution and anticipation as possible.

CHEMICAL REACTIONS

Concrete damage can occur when chemical reactions are present. It can be surprising how small an amount of chemicals can do serious structural damage to concrete. To expand on this, let's take a look at some examples of chemical reactions and how they affect concrete.

Acid

Most people know that acid can have serious reactions with a number of materials. Concrete can also be affected by acid exposure. When acid attacks concrete, it

doi: 10.1016/B978-1-85617-549-4.00003-4

concentrates on the products of hydration of cement. For example, calcium silicate hydrate can be adversely affected by exposure to acid. Sulfuric acid works to weaken concrete, and if it is able to reach the steel reinforcing members, the steel can be compromised. All of this contributes to a failing concrete structure.

Visual inspections may reveal a loss of cement paste and aggregate from the matrix. Cracking, spalling, and discoloration can be expected when acid deteriorates steel reinforcements. Laboratory analysis may be needed to indentify the type of chemical causing the damage.

It is a code violation to embed aluminum conduits and pipes in concrete, unless the aluminum is coated or covered to prevent aluminum-concrete reaction or electrolytic action between aluminum and steel.

Table 3.1 Causes of Distress and Deterioration of Concrete

Causes of Distress and Deterioration of Concrete
Accidental Loadings
Chemical Reactions Acid attack Aggressive-water attack Alkali-carbonate rock reaction Alkali-sllica reaction Miscellaneous chemical attack Sulfate attack Construction Errors Corrosion of Embedded Metals
Design Errors Inadequate structural design Poor design details
Erosion Abrasion Cavitation
Freezing and Thawing
Settlement and Movement Shrinkage Plastic Drying
Temperature Changes Internally generated Externally generated Fire
Weathering

Courtesy of United States Army Corps of Engineers

How can you create a more defensive concrete where chemical reactions are anticipated? Portland-cement concrete does not fare well when exposed to acid. When faced with this type of concrete, an approved coating or treatment is about the best you can do. Using a dense concrete with a low water-cement (w/c) ratio can provide acceptable protection against mild acid exposure.

Aggressive-water attack

An aggressive-water attack has to do with water that has a low concentration of dissolved minerals. Soft water is aggressive water, and it will leach calcium from cement paste or aggregates. This is not common in the United States. When this type of attack occurs, it is a slow process. The danger is greater in flowing waters. This is due to a fresh supply of aggressive water coming into contact with the concrete.

If you conduct a visual inspection and find rough concrete where the paste has been leached away, it could be an aggressive-water defect. Water can be tested to determine if the water quality is responsible for the damage. When testing indicates that water may create problems prior to construction, a cement-based (not Portland) coating can be applied to the exposed concrete structures.

When conduits are installed in concrete, the diameter of the conduit should not be more than one-third of the overall thickness of the concrete slab where it is being installed.

Alkali-carbonate rock reaction

Alkali-carbonate rock reaction can result in damage to concrete, but it can also be beneficial. Our focus is on the destructive side of this action. This occurs when impure dolomitic aggregates exist. When this type of damage occurs, you are likely to find map or pattern cracking, and the concrete will look as if is it swelling.

Alkali-carbonate rock reaction differs from alkali-silica reaction in that there is a lack of silica gel exudations at cracks. Petrographic examination can be used to confirm the presence of alkali-carbonate rock reaction. To prevent this type of problem, contractors should avoid using aggregates that are, or are suspected to be, reactive.

Conduits embedded in concrete must be spaced a minimum distance that would equal not less than three times the diameter of the conduit being installed.

Alkali-silica reaction

An alkali-silica reaction can occur when aggregates containing silica that is soluble in highly alkaline solutions may react to form a solid, nonexpansive, calcium-alkali-silica complex or an alkali-silica complex that can imbibe considerable amounts of water and expand. This can be disruptive to concrete.

Concrete that shows map or pattern cracking and a general appearance of swelling could be a result of an alkali-silica reaction. Using concrete that contains less than 0.60 percent alkalies can allow pattern cracking to occur.

Various chemical attacks

Concrete is fairly resistant to chemical attack. For a substantial chemical attack to have degrading effects of a measurable nature, a high concentration of chemical is required. Solid, dry chemicals are rarely a risk to concrete. Chemicals that are circulated in contact with concrete do the most damage.

When concrete is subjected to aggressive solutions under positive differential pressure, the concrete is particularly vulnerable. The pressure can force aggressive solutions into the matrix. Any concentration of salt can create problems for concrete structures. Temperature plays a role in concrete destruction with some chemical attacks. Dense concrete that has a low w/c provides the greatest resistance. The application of an approved coating is another potential option for avoiding various chemical attacks.

> When concrete joints are created, they must be located in a manner that will not have an adverse effect on the strength of the concrete in which they are installed.

Sulfate attack

A sulfate attack on concrete can occur from naturally occurring sulfates of sodium, potassium, calcium, or magnesium. These elements can be found in soil or in groundwater. Sulfate ions in solution will attack concrete. Free calcium hydroxide reacts with sulfate to form calcium sulfate, also known as gypsum. When the gypsum combines with hydrated calcium aluminate to form calcium sulfoaluminate, either reaction can result in an increase in volume. Additionally, a purely physical phenomenon in which the growth of crystals of sulfate salts disrupts the concrete can occur. Map and pattern cracking and general disintegration of the concrete are signs of a sulfate attack.

Preventing sulfate attacks can usually be done with the use of a dense, high-quality concrete that has a low water-cement ratio. A Type V or Type II cement is a good choice. If pozzolan is used, a laboratory evaluation should be done to establish the expected improvement in performance.

> Unless otherwise authorized, all bending of reinforcement material for concrete should be done while the material is cold.

Table 3.2 Relating Symptoms to Causes of Distress and Deterioration of Concrete

Causes	Symptoms							
	Construction Faults	Cracking	Disintegration	Distortion/Movement	Erosion	Joint Failures	Seepage	Spalling
Accidental Loadings		X						X
Chemical Reactions		X	X				X	
Construction Errors	X	X				X	X	X
Corrosion		X				X	X	X
Design Errors		X						X
Erosion			X		X			
Freezing and Thawing		X	X					X
Settlement and Movement		X		X		X		X
Shrinkage	X	X		X				
Temperature Changes		X				X		X

Courtesy of United States Army Corps of Engineers

Poor workmanship

Poor workmanship accounts for a number of concrete issues. It is simple enough to follow proper procedures, but there are always times when good practices are not used. The best solution to poor workmanship is to prevent it in the first place. Unfortunately, this is much easier to say than to do. All sorts of problems can occur when quality workmanship is not ensured. The following are some of the key causes for such problems:

- Adding too much water to concrete mixtures
- Poor alignment of formwork
- Improper consolidation
- Improper curing
- Improper location and installation of reinforcing steel members
- Movement of form work
- Premature removal of shores or reshores
- Settling of concrete
- Settling of subgrade
- Vibration of freshly placed concrete
- Adding water to the surface of fresh concrete
- Miscalculating the timing for finishing concrete
- Adding a layer of concrete to an existing surface
- Use of a tamper
- Jointing

CORROSION

Corrosion of steel reinforcing members is a common cause of damage to concrete. Rust staining will often be present during a visual inspection if corrosion is at work. Cracks in concrete can tell a story. If they are running in straight lines, as parallel lines at uniform intervals that correspond with the spacing of steel reinforcement materials, you can suspect corrosion is at the root of the problem. In time, spalling will occur. Eventually, the reinforcing material will become exposed to a visual inspection.

> Unless otherwise authorized, it is a code violation to weld crossing bars that will reinforce concrete.

Techniques for stopping, or controlling, corrosion include the use of concrete with low permeability. In addition, good workmanship is needed. Here are some tips to follow:

- Use as low a concrete slump as practical.
- Cure the concrete properly.
- Provide adequate concrete cover over reinforcing material.

- Provide suitable drainage.
- Limit chlorides in the concrete mixture.
- Pay special attention to any protrusions, such as bolts and anchors.

Design mistakes

Design mistakes are divided into two categories: those that are a result of inadequate structural design and those that are a result of a lack of attention to relatively minor design details. In the case of structural design errors, the result can be anticipated. It will generally result in a structural failure.

Identifying structural design mistakes involves two types of symptoms. Spalling indicates excessively high compressive stress. Cracking and spalling can also indicate high torsion or shear stresses. High tensile stresses will cause cracks. Petrographic analysis and strength testing of concrete is required if any of the concrete elements are to be reused after such failures. The best prevention requires careful attention to detail. Design calculations should be checked thoroughly. Flaws in design details account for most of these types of problems. The following are some examples of design factors to consider:

- Poor design details
- Abrupt changes in section
- Insufficient reinforcement at reentrant corners and openings
- Inadequate provision for deflection
- Inadequate provision for drainage
- Insufficient travel in expansion joints
- Incompatibility of materials
- Neglect of creep effect
- Rigid joints between precast units
- Unanticipated shear stresses in piers, columns, or abutments
- Inadequate joint spacing in slabs

Abrasion

Abrasion damage can occur from waterborne debris. The debris typically rolls and grinds against concrete when it is in the water and in contact with concrete structures. Spillway aprons, stilling basin slabs, and lock culverts and laterals are the most likely types of structures to be affected by abrasion. This is usually a result of poor hydraulic design. Another cause for abrasion can be a boat hull hitting a concrete structure.

> When groups of reinforcing bars are bundled together as a reinforcement device for concrete, the bundle must not contain more than four bars.

When abrasion is in play, concrete structures tend to wind up with a smooth surface. Long, shallow grooves in a concrete surface and spalling along monolith joints indicate abrasion. The three major factors in avoiding abrasion damage are design, operation, and materials. Here are some tips to keep in mind:

- Use hydraulic model studies to test designs.
- A 45-degree fillet installed on the upstream side of the end sill has resulted in a self-cleaning stilling basin.
- Recessing monolith joints in lock walls and guide walls will minimize stilling basin spalling caused by barge impact and abrasion.
- Balanced flows should be maintained into basins by using all gates to avoid discharge conditions where eddy action is prevalent.
- Periodic inspections are needed to locate the presence of debris.
- Basins should be cleaned periodically.
- All materials used must be tested and evaluated.
- Install abrasion-resistant concrete.
- Fiber-reinforced concrete should not be used for repairing stilling basins or other hydraulic structures that are subject to abrasion.
- Coatings that produce good results against abrasion include polyurethanes, epoxy-resin mortar, furan-resin mortar, acrylic mortar, and iron aggregate toppings.

Cavitation

Cavitation-erosion is a result of complex flow characteristics of water over concrete surfaces. For damage to occur, the rate of water flow normally has to exceed 40 feet per second. Fast water and irregular surface areas of concrete can result in cavitation. Now we get to the interesting part: The surface irregularity and water speed create bubbles. The bubbles are carried downstream and have a lowered vapor pressure. Once the bubbles reach a stretch of water that has normal pressure, the bubbles collapse. The collapse is an implosion that creates a shock wave. Once the shock wave reaches a concrete surface, the wave causes a very high stress over a small area. When this process is repeated, pitting can occur. This type of cavitation has affected concrete spillways and outlet works of many high dams. Prevention has to do with design, materials, and construction practices. The following list highlights some of the key considerations:

- Include aeration in a hydraulic design.
- Use concrete designed with low w/c.
- Use hard, dense aggregate particles.
- Steel-fiber concrete and polymer concrete can aid in the fight against cavitation.
- Neoprene and polyurethane coatings can assist in the fight against cavitation. However, coatings are rarely used because they might prevent the best adhesion to concrete. Any rip or tear in the coating can cause a complete stripping of the coating over time.
- Maintain approved construction practices.

Bundle reinforcing bars must be enclosed within either stirrups or ties.

FREEZING AND THAWING

A pattern of freezing and thawing during the curing of concrete is a serious concern. Each time the concrete freezes, it expands. Hydraulic structures are especially vulnerable to this type of damage. Fluctuating water levels and underspraying conditions increase the risk. Using deicing chemicals can accelerate damage to concrete. It will cause pitting and scaling. Core samples are likely to be needed to assess the damage.

Prevention is the best cure. Provide adequate drainage, where possible. Work with low-w/c concrete. Use adequate entrained air to provide suitable air-void systems in the concrete. Select aggregates best suited for the application. Make sure that the concrete cures properly.

SETTLEMENT AND MOVEMENT

Settlement and movement can be the result of differential movement or subsidence. Concrete is rigid and cannot stand much differential movement. When it occurs, stress cracks and spall are likely to occur. Subsidence causes entire structures, or single elements of entire structures, to move. If subsidence occurs, the concern is not cracking or spalling. The big risk is stability against overturning or sliding.

A failure via subsidence is generally related to a faulty foundation. Long-term consolidations, new loading conditions, and related faults are contributors to subsidence. Geotechnical investigations are often needed when subsidence is evident.

Things to look for when structure movement is suspected include cracking, spalling, misaligned members, and water leakage. Specialists are normally needed for these types of investigations.

Spiral reinforcement for cast-in-place concrete must not be less than ⅜ inch in diameter.

SHRINKAGE

Shrinkage occurs when concrete is deficient in its moisture content. The shrinkage can occur while the concrete is setting or after it is set. When the condition occurs during setting, it is called plastic shrinkage. Drying shrinkage happens after the concrete is set.

Plastic shrinkage is associated with bleeding, which is the appearance of moisture on the surface of concrete. This is usually caused by the settling of heavier components in a mixture. Bleed water typically evaporates slowly from the surface of concrete. When evaporation is occurring faster than water is being supplied to the surface by bleeding, high-tensile stresses can develop. The stress can lead to cracks on the concrete surface.

Cracks caused by plastic shrinkage usually occur within a few hours of concrete placement. The cracks are normally isolated. They also tend to be wide and shallow. Pattern cracks are not generally caused by plastic shrinkage.

> Spacing requirements for shrinkage and temperature reinforcement must be spaced not farther apart than five times a slab's thickness and not farther apart than 18 inches.

Weather conditions contribute to plastic shrinkage. If the conditions are expected to be conducive to plastic shrinkage, protect the pour site. This can be done with windbreaks, tarps, and similar arrangements to prevent excessive evaporation. In the event the early cracks are discovered, revibration and refinishing can solve the immediate problem.

Drying shrinkage is a long-term change in volume of concrete caused by the loss of moisture. A combination of this shrinkage and restraints will cause tensile stresses and lead to cracking. The cracks will be fine, and there will be no indication of movement. The cracks are typically shallow and only a few inches apart. Look for a blocky pattern to the cracks. The identification can be confused with thermally induced deep cracking that occurs when dimensional change is restrained in newly placed concrete by rigid foundations or by old lifts of concrete.

To reduce drying shrinkage, do the following:

- Use less water in concrete.
- Use larger aggregate to minimize paste content.
- Use a low temperature to cure concrete.
- Dampen the subgrade and the concrete forms.
- Dampen aggregate if it is dry and absorbent.
- Provide adequate reinforcement.
- Provide adequate contraction joints.

TEMPERATURE CHANGES

Temperature changes can affect shrinkage. The heat of hydration of cement in large placements can present problems. Climatic conditions involving heat also have the capability to affect concrete. Fire damage, while rare, can also contribute to problems associated with excessive heat.

> The code allows one to assume that the ends of columns built integrally with a structure will remain fixed. This assumption comes into play when one is computing gravity load moments on columns.

Did you know that hydration of concrete can raise the temperature of freshly placed concrete by up to 100°F? It can! Rarely is the temperature increase consistent in all of the concrete, and this can generate problems. Cracks can occur. The cracks should be shallow and isolated. To avoid this, do the following:

- Use low-heat cement.
- Pour concrete at the lowest reasonable temperature.
- Select aggregates with low moduli of elasticity and low coefficients of thermal expansion.

External temperature changes can result in cracking that will appear as regularly spaced cracks. There may be spalling at expansion joints. Using contraction and expansion joints can help prevent this damage.

There are many potential causes for concrete failure. Extended education, experience, and scientific testing are often required to identify clearly the causes of failure. There is always more to learn. Keeping an open mind and immersing yourself in the components of concrete are the best ways to achieve success.

CHAPTER

Planning and design of concrete repair

4

Sometimes concrete needs to be repaired. A successful repair relies on many factors. One must consider the best repair strategy, material, and procedure. The best attempt at a repair is likely to fail if the wrong material is used for the repair.

When you are planning a concrete repair, you must consider various options and evaluate the best type of material to implement in the repair or rehabilitation of concrete structures. You may find the choices for repair to be controversial in some venues. Appropriate experts may need to be consulted for ensured success. There are, however, many facts that are proven and can be trusted. They are what we are going to deal with in this chapter.

COMPRESSIVE STRENGTH

How much compressive strength is needed in material for concrete repairs? If it is determined that the existing concrete structure is of adequate compressive strength, then the repair material should be of a similar compressive strength. There are few instances where beefing up the compressive strength in a repair is beneficial. An exception is the repair of concrete that is damaged by erosion. When erosion is at fault for a defect, using a higher compressive strength is a valid decision.

T-beam construction requires that the flange and web must be built integrally or otherwise effectively bonded together.

MODULUS OF ELASTICITY

Modulus of elasticity is a measure of stiffness with higher-modulus materials exhibiting less deformation under load compared to low-modulus materials. When making

doi: 10.1016/B978-1-85617-549-4.00004-6

a repair, the modulus of elasticity should be similar to that of the concrete substrate. This allows for uniform load transfer across a repaired section. Using materials with a lower modulus of elasticity will exhibit lower internal stresses. This reduces the potential for cracking and delamination of a repair.

> Clear spacing between ribs in joist construction cannot be more than 30 inches.

THERMAL EXPANSION

Thermal expansion is going to happen with concrete. If a polymer is used as a repair material, the result will often be cracking, spalling, or delamination of the repair. The coefficient of thermal expansion must be considered for suitable repair materials. When you compare a polymer to concrete, the coefficient of thermal expansion for a polymer is likely to be up to 14 times greater than that of concrete. Large repairs and overlays are especially vulnerable to cracking due to thermal expansion.

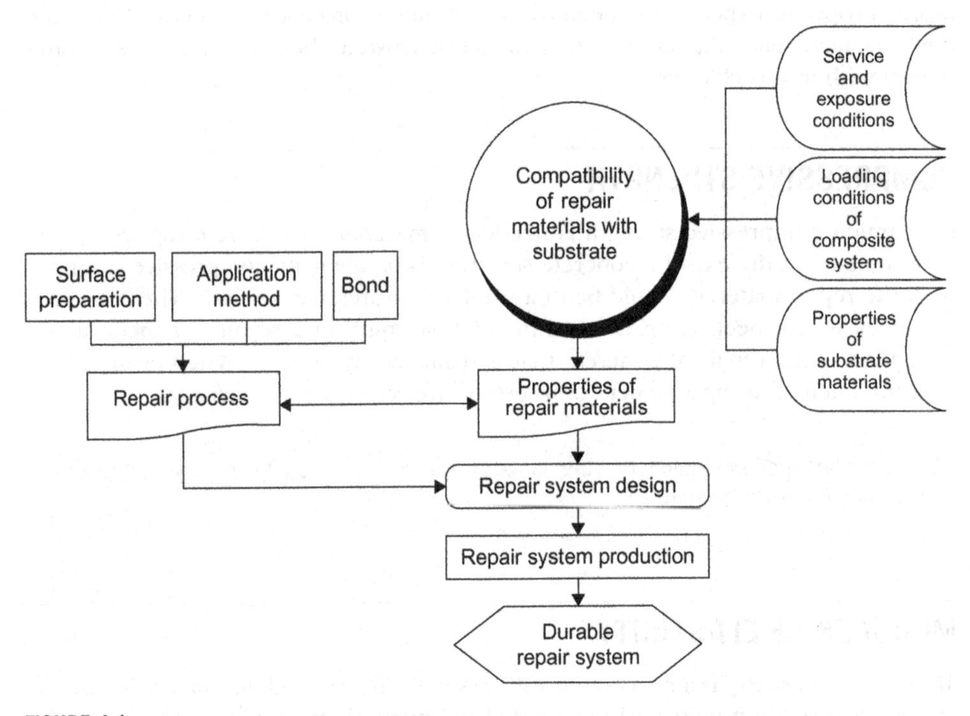

FIGURE 4.1

Factors affecting the durability of concrete repair systems. (From Emmons and Vaysburd, 1995)
Courtesy of United States Army Corps of Engineers

Slab thickness in joist construction must not be less than 2 inches in depth.

BONDING

The bonding between repair material and concrete is a key element in a successful repair. For concrete to accept a good bond with a repair, the concrete should be properly prepared. Polymer adhesives provide a better bond of plastic concrete to hardened concrete than can be obtained with a cement slurry or the plastic concrete alone. This creates some controversy. Many experienced people indicate that polymer bonds are less than 25 percent better than properly prepared concrete surfaces without adhesives.

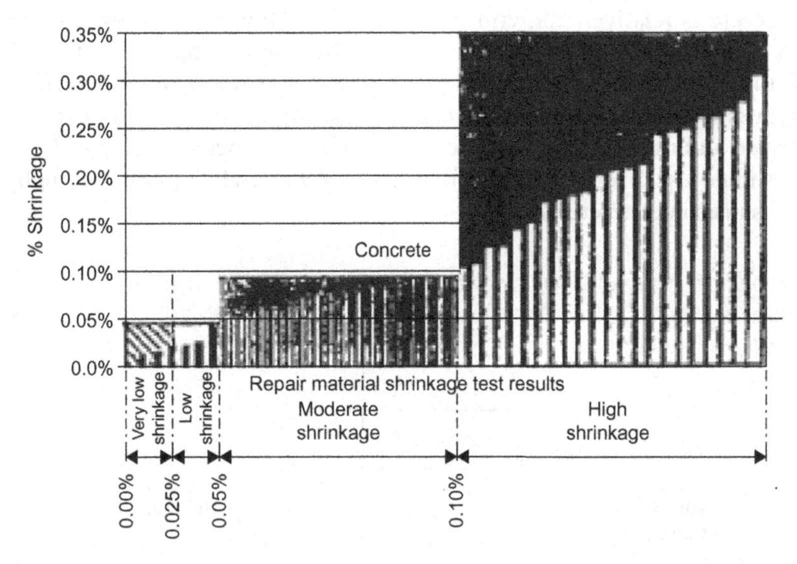

FIGURE 4.2

Classification of repair materials based on drying shrinkage. (From Emmons and Vaysburd, 1995) *Courtesy of United States Army Corps of Engineers*

DRYING SHRINKAGE

Drying shrinkage is a concern in concrete repairs. Existing concrete that is in need of repair is unlikely to shrink. However, patches and repairs that are made fresh are subject to shrinking, and this can compromise the repair. To avoid shrinkage, the material used for repairing old concrete should be made with a low w/c ratio. There are repair products available that offer minimum shrinkage. The goal is to use a material that will provide minimum shrinkage.

When estimating the thickness of concrete to be applied as the minimum amount required to cover material, the floor finish may be counted for nonstructural considerations.

CREEP

Repair materials should have a creep factor consistent with the material being repaired. Stress relaxation through tensile creep reduces the potential for cracking. Refer to the manufacturer's documents when selecting an appropriate creep rate for various repairs.

PERMEABILITY

Good concrete is relatively impermeable to liquids. However, moisture evaporates at a surface, and replacement liquid is pulled to the evaporating surface by diffusion. This has to be considered when making a repair. Any large patch or overlay that is made with an impermeable material can trap moisture between the existing concrete surface and the seal made by the repair. If this happens, the repair is likely to fail. You should use a repair material that has low water absorption and high water vapor transmission characteristics.

The minimum number of longitudinal bars in compression members is four bars within rectangular or circular ties.

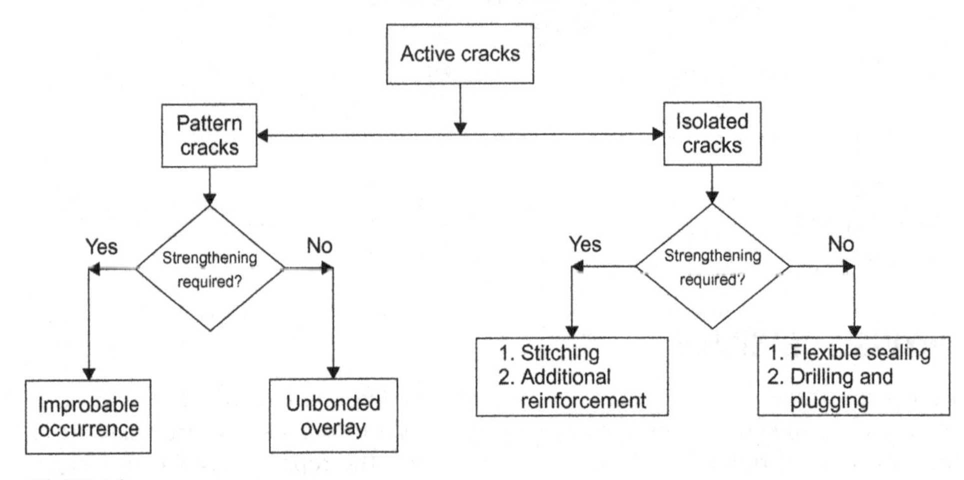

FIGURE 4.3

Selection of repair method for active cracks. (From Johnson, 1965)
Courtesy of United States Army Corps of Engineers

PLANNING A REPAIR

Planning a concrete repair requires consideration of many factors, including the following:

- Application conditions
- Geometry
- Temperature
- Moisture
- Location
- Service conditions
- Downtime
- Traffic
- Temperature
- Chemical attack
- Appearance
- Service life

Depth and orientation of a concrete repair are important considerations. Thick sections have heat generated during curing of some repair materials. Thermal stress can

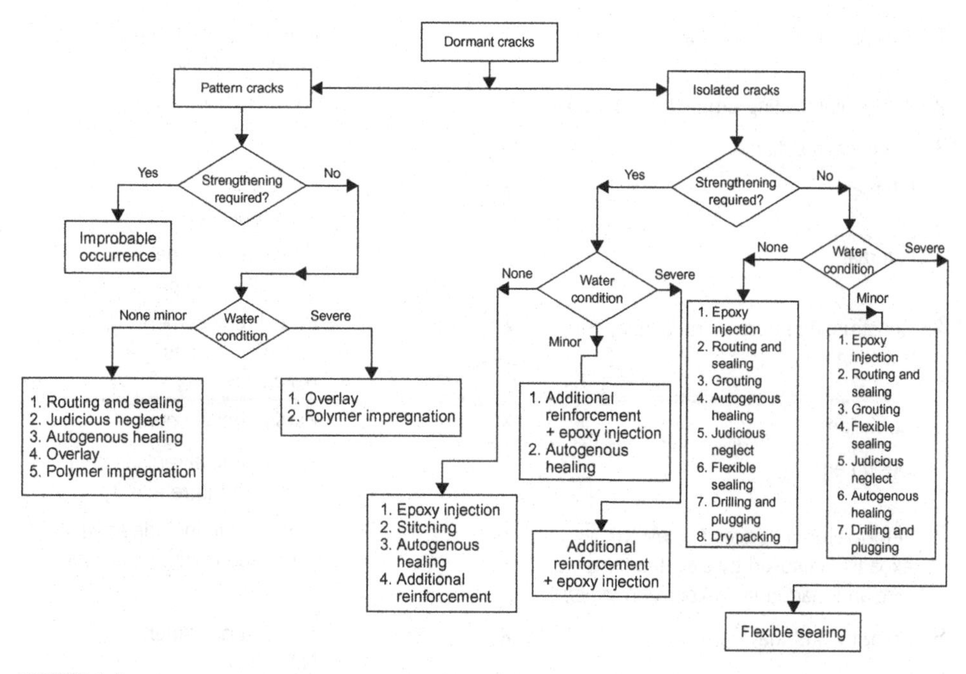

FIGURE 4.4

Selection of repair method for dormant cracks. (From Johnson, 1965)
Courtesy of United States Army Corps of Engineers

occur that is beyond an acceptable level. Shrinkage is another concern. Thin layers of concrete used as a repair are subject to spalling. An advantage to polymer materials is that they can be used in thin layers. Aggregate size is determined by the thickness of a repair. Repairs made overhead must be made in such a way that they will not sag.

The minimum number of longitudinal bars in compression members is three bars within triangular ties.

Portland cement hydration stops at, or near, freezing (32°F). Latex emulsions do not coalesce to form films at temperatures below about 45°F. Materials that can be used in colder temperatures generally require longer setting times. In contrast, high temperatures may make a repair material set faster and result in a decrease in the working life of the material.

Table 4.1 Causes and Repair Approaches for Spalling and Disintegration

Cause	Deterioration likely to Continue		Repair Approach
	Yes	***No***	
1. Erosion (abrasion, cavitation)	X		Partial replacement Surface coatings
2. Accidental loading (impact, earthquake)		X	Partial replacement
3. Chemical reactions			
Internal	X		No action Total replacement
External	X	X	Partial replacement Surface coatings
4. Construction errors (compaction, curing, finishing)	X		Partial replacement Surface coatings No action
5. Corrosion	X		Partial replacement
6. Design errors	X	X	Partial or total replacement based on future activity
7. Temperature changes (excessive expansion caused by elevated temperature and Inadequate expansion joints)	X		Redesign to include adequate joints and partial replacement
8. Freezing and thawing	X		Partial replacement No action

NOTE: This table is Intended to serve as a general guide only. It should be recognized that there are probably exceptions to all of the items listed.
Courtesy of United States Army Corps of Engineers

Having water come into contact with fresh concrete is not acceptable in most repairs. Grouting, external waterproofing, or diversion systems are commonly used to prevent interference from moving water while working with concrete. If you are working with polymers, some of them will not adhere in moist conditions. On the other hand, some polymers are not affected by moisture.

> The minimum number of longitudinal bars in compression members is six bars enclosed by spirals.

Table 4.2 Repair Methods for Spalling and Disintegration

Repair Approach	Repair Method
1. No action	Judicious neglect
2. Partial replacement (replacement of only damaged concrete)	Conventional concrete placement Drypacking Jackeling Preplaced-aggregate concrete Polymer impregnation Overlay Shotcrete Underwater placement High-strength concrete
3. Surface coating	Coatings Overlays
4. Total replacement of structure	Remove and replace

NOTE: Individual repair methods are discussed in Chapter 6, except those for surface coatings, which are discussed In Chapter 7.
Courtesy of United States Army Corps of Engineers

The location of a needed repair can have an impact on both materials and procedures. Some locations limit the type of equipment that can be used. Some repair materials are odorous, toxic, or combustible. All of these factors must be considered when planning a repair.

Sometimes materials that will set quickly are needed to reduce downtime. In the case of heavy vehicular traffic, repair materials need to have a high strength rating and good abrasion and skid resistance. You also have to consider concrete deterioration.

High-service temperatures can affect the performance of some polymers. While polymers can be sensitive to solvents, most polymers resist most acids and sulfates. Soft water can damage Portland cement products. Matching patches and repairs can be very difficult. If appearance is a factor, you will have to do your research to come up with a close, visual match. You must also keep in mind how long will the repair be expected to last when you are choosing the type of repair material to use.

MANUFACTURER'S DATA

Most manufacturers offer information on the qualities of their products. You should be able to find information on (1) the compressive strength, (2) the tensile strength, (3) the slant-shear bond, and (4) the modulus of elasticity.

Even with this information, you may have more questions that are not answered in the manufacturer's documents—for example, drying shrinkage, tensile bond strength, creep, absorption, and water vapor transmission. If you are not provided with this information in the planning phase, request it from the manufacturers.

The planning phase is critical to a successful outcome. You may have to spend a few hours on research, but the result should be a better repair. Don't guess. Make your decisions based on facts.

CHAPTER

Concrete removal and preparation for repair

5

It is very common for some concrete to be removed before making a repair to a concrete structure. Many jobs require existing concrete to be prepared properly to receive a repair. When the damage done to concrete does not threaten the structural integrity of a structure, the concrete will probably not have to be removed.

Environmental impact from concrete removal is often an issue. This is especially true when the work is done near a waterway. In some cases, concrete debris can be allowed to enter a waterway because some of the aggregate used in concrete is natural river gravel anyway. Therefore, allowing it to fall into the water is essentially returning the gravel from its place of origin. Another possibility, when allowed, is to dispose of large pieces of concrete in water so it can create an artificial reef for fish. When this is an acceptable means of disposal, it reduces the cost of deconstruction considerably.

When concrete is removed, it is done in such a way that sound concrete can be reached to base a repair on. It is not uncommon for multiple methods to be used on a single structure to remove concrete. For example, a presplitting method might be used to weaken concrete so an impacting method can be used to complete the concrete removal. Another example could be the cutting of a section of concrete to delineate an area in which an impacting method will be used for concrete removal.

When demolishing concrete, you must be aware of any utilities or other embedded items in the existing concrete. A review of as-built drawings can render assistance in projecting obstacle locations.

doi: 10.1016/B978-1-85617-549-4.00005-8

Table 5.1 A General Classification of Concrete Removal Methods Applicable for Concrete Repair

Category	Description	Specific Methods
Blasting	Blasting methods employ rapidly expanding gas confined within a series of boreholes to produce controlled fracture and removal of concrete	Explosive blasting
Crushing	Crushing methods employ hydraulically powered jaws to crush and remove the concrete	Mechanical crushing, boom-mounted Mechanical crushing, portable
Cutting	Cutting methods employ full-depth perimeter cuts to disjoint concrete for removal as a unit or units	Abrasive-water-jet cutting Diamond-blade cutting Diamond-wire cutting Stitch driing Thermal cutting
Impacting	Impacting methods employ repeated striking of the surface with a mass to fracture and spall the concrete	Mechanical impacting, hand-held Mechanical impacting, boom-mounted Mechanical impacting, spring-action
Milling	Milling methods generally employ abrasion or cavitation-erosion techniques to remove concrete from surfaces	Hydromilling Rotary head milling
Presplitting	Presplitting methods employ wedging forces in a designed pattern of boreholes to produce a controlled cracking of the concrete to facilitate removal of concrete by other means	Presplitting, chemical-expansive agents Presplitting, piston-jack splitter, Presplitting, plug-and-feather splitter

Courtesy of the United States Army Corps of Engineers

REMOVAL METHODS

Several methods can be used to remove concrete:

- Blasting
- Crushing
- Cutting
- Impacting
- Milling
- Presplitting

Table 5.2 Selection Features and Considerations for Concrete Removal Methods

Category	Method	Features	Considerations
Blasting	Explosive blasting	Method applicable for removal from mass concrete structures Method is most expedient and, in many cases, the most cost-effective means of removing large volumes where 250 mm (10 in.) or more of face is to be removed Produces reasonably small size debris that is easily handled	Requires highly skilled personnel for design and execution of blasting plan Stringent safely regulations must be complied with regard to the transportation, storage, and use of explosives because of their inherent dangers Sequential blasting techniques must be employed to reduce peak blast energies and, thereby, limit damage to surrounding property resulting from air-blast pressure, ground vibration, and fly rock Control blasting techniques should be employed to limit damage to concrete that remains
Crushing	Mechanical crushing, boom-mounted	Method applicable for removing concrete from decks, walls, columns, and other concrete members where shearing plane depth is 1.8 m (6 ft) or less Boom allows removal from vertical and overhead members Steel reinforcing can be cut Limited noise and vibration is produced Pulverizing jaw attachment can debond the concrete from the steel reinforcement for purpose of recycling both Method produces relatively small debris that is easily handled	Method is more applicable for total demolition of a concrete member than for removal to rehabilitate or repair Boundaries must be saw cut to limit overbreakage Removal must be started from a free edge or a hole cut in member Exposed reinforcing steel is damaged beyond reuse Production rates vary depending on condition of concrete and amount of reinforcement

(*Continued*)

Table 5.2 (Continued)

Category	Method	Features	Considerations
	Mechanical crushing, portable	Method applicable for removal from decks, walls, and other members where shearing plane depth is 300 mm (12 in.) or less Method can be used to remove concrete in areas of limited work space Llimited noise and vibration is produced Produces small size debris that is easily handled	Requires two men to handle (weighs approximately 45 kg (100 lb)) Reinforcing steel is damaged beyond reuse Crushing must be started from a free edge or a hole cut in member Boundaries must be saw cut to limit overbreakage Production rates are low
Cutting	Abrasive-water-jet cutting	Method applicable for making cutouts through slabs, walls, and other concrete members where access to only one face is feasible and depth of cut is 500 mm (20 in.) or less Abrasives enable jet to cut steel reinforcing and hard aggregates Irregular and curved cutouts can be made Cutouts can be made without overcutting corners Cuts can be made flush with adjoining members No heat, vibration, or dust is produced Handling of debris is more efficient as bulk of concrete is removed as units	Cutting is typically slower and more costly than diamond-blade sawing Controlling flow of waste water may be required Personnel must wear hearing protection because of the high levels of noise produced Additional safety precautions are required because of the high levels of noise produced Additional safety precautions are required because of high waler pressures (200-340 MPa (30,000-50,000 psi) produced by system

Diamond-blade cutting	Method applicable for making cutouts through slabs, walls, and other concrete members where access to only one face is feasible and depth of cut is 600 mm (24 in.) or less Precision cuts can be made No dust or vibration is produced Handling of debris is more efficient as bulk of concrete is removed as units	Selection of the type diamonds and metal bond used in blade segments is based on the type (hardness) and percent of coarse aggregate and on the percent of steel reinforcing in cut The higher the percent of steel reinforcement in cuts, the slower and more costly the cutting The harder the aggregate, the slower and more costly the cutting Controlling flow of waste water may be required Special blades with flush-cut arbors are required to make cuts flush with adjoining members
Diamond-wire cutting	Method applicable for making cutouts through concrete where depth of cut is greater than can be economically cut with the diamond-blade saw Cuts can be made through mass concrete and in areas of difficult access Overcutting of corner can be avoided if cut started from drilled hole at corner No dust or vibration is produced Handling of debris is more efficient as bulk of concrete is removed as units	The wire saw is a specialty tool that for many jobs will not be as cost effective as other techniques, such as blasting, impacting, and presplitting Selection of type diamonds and metal bond used in beads is based on type (hardness) and percent of coarse aggregate and percent of steel reinforcing in cut The higher the percent of steel reinforcement in cuts, the slower and more costly the cutting The harder the aggregate, the slower and more costly the cutting Beads with embedded diamonds last longer, but are more expensive than beads with electroplated diamonds (single layer)

(*Continued*)

Table 5.2 (Continued)

Category	Method	Features	Considerations
			Wires with beads having embedded diamonds should be of sufficient length to complete cut as replacement will not fit into cut (wear reduces wire diameter and, thereby, cut opening as cutting proceeds) Deep cutouts that are formed by three or more boundary cuts may require tapering to avoid binding during removal Controlling flow of waste water may be required
	Stitch drilling	Method applicable for making cutouts through concrete members where access to only one face is feasible and depth of cut is greater than can be economically cut by diamond-blade saw Handling of debris is more efficient as bulk of concrete is removed as units	Rotary-percussion drilling is significantly more expedient and economical than diamond-core for nonreinforced concrete Diamond-core drilling is more applicable than rotary-percussion drilling for reinforced concrete The greater the percentage of steel reinforcement contained within a cut, the slower and more costly the cutting Depth of cuts is dependent on accuracy of drilling equipment in maintaining overlap between holes with depth and on the diameter of boreholes drilled The deeper the cut, the greater borehole diameter required to maintain overlap between adjacent holes and the greater the cost Uncut portions between adjacent boreholes will prevent removal

			Concrete toughness for percussion drilling and aggregate hardness for diamond coring will affect cutting rate and cost Personnel must wear hearing protection because of the high levels of noise produced
	Thermal cutting	Method applicable for making cutouts through heavily reinforced decks, beams, walls, and other reinforced members where site conditions allow efficient flow of molten concrete from cuts Method is an effective means of cutting prestressed members Irregular shapes can be cut Minimal vibration and dust produced Handling of debris is more efficient as bulk of concrete is removed as units	Method is of limited commercial availability and is costly Remaining concrete has thermal damage with more extensive damage occurring around steel reinforcement Noise, smoke, and fumes are produced Personnel must be protected from heat and hot flying rock produced by cutting operation Additional safety precautions are required because of hazards associated with storage handling, and use of compressed and flammable gases
Impacting	Mechanical impacting boom-mounted breaker	Method is applicable for both full and partial depth removals where required production rates are greater than can be economically achieved by the use of hand-held breakers Boom allows concrete to be removed from vertical and overhead members Boom-mounted breakers are widely available commercially Method produces easily handled debris	The blow energy delivered to the concrete should be limited to protect the structure being repaired and surrounding structures from damage resulting from the high cyclic energy generated Performance is a function of concrete soundness and toughness Productivity is significantly reduced when boom is operated from top of wall because of the operator's limited view of the removal operation Care must be taken to avoid damage to supporting members Concrete that remains may be damaged (microcracking) along with reinforcing steel

(*Continued*)

Table 5.2 (Continued)

Category	Method	Features	Considerations
			Saw cuts at boundaries should be employed to reduce the occurrence of feathered edges Dust is produced Personnel must wear hearing protection because of the high levels of noise produced
	Mechanical impacting hand-held breaker	Method is applicable for work involving limited volumes of concrete removal and for removal in areas of limited access Hand-held breakers are widely available commercially Breakers can be operated by unskilled labor Method produces relatively small debris that is easily handled	Hand-held breakers are generally not applicable for large volumes of removal, except where blow energy must be limited Performance is a function of concrete soundness and toughness Significant loss in productivity occurs when breaking action is other than downward Removal boundaries will likely require 25-min (1-in) deep or greater saw cut to reduce the occurrence of feathered edges Concrete that remains may be damaged (microcracking) Size of breakers for bridge decks is typically limited to 14-kg (30-lb) class for removal above reinforcement and 7-kg (15-lb) class from around reinforcement Dust is produced Personnel must wear hearing protection because of the high levels of noise produced

	Mechanical impacting spring-action hammer	Method is applicable for breaking concrete pavement, decks, walls, and other thin members where production rates required are greater than can be economically achieved by the use of handheld breakers For decks, hammer can completely punch through slab with each blow leaving only the reinforcing steel Method produces easily handled debris	Method is more applicable for total demolition of a concrete member than for removal to rehabilitate or repair The blow energy delivered to the concrete should be limited to protect the structure being repaired and surrounding structures from damage resulting from the high cyclic energy generated Care must be taken to avoid damage to supporting members Performance is a function of concrete soundness and toughness Concrete that remains may be damaged (microcracking) along with reinforcing steel Saw cuts at boundaries should be employed to reduce the occurrence of feathered edges
Milling	Hydromilling (Also known as hydrodemolition anc water-jet blasting)	Method is applicable for removal of deteriorated concrete from surfaces of decks and walls where removal depth is 150 mm (6 in.) or less Method does not damage the concrete that remains Steel reinforcing is left undamaged for reuse Method produces easily handled, aggregate-size debris	Method is costly Productivity is significantly reduced when sound concrete is being removed Removal profile will vary with changes in depth of deterioration Holes through member (blowouts) are a common occurrence when removal is near full depth of member Repair of blowouts requires additional material and form work, thereby increasing repair time and cost Method requires large sources of potable water (the water demand for some units exceeds 4,000 L/hr (1,000 gal/hr))

(*Continued*)

Table 5.2 (Continued)

Category	Method	Features	Considerations
			Laitence coating that is deposited on remaining surfaces during removal should be washed from surface before coating dries Flow of waste water may have to be controlled An environmental impact statement will be required if waste water is to enter a waterway Personnel must wear hearing protection because of the high level of noise produced Fly rock is produced Additional safety requirements are required because of the high pressures (100–300-MPa (16,000–40,000-psi) range) produced by the system
	Rotary-head milling	Method is applicable for removing deteriorated concrete from mass structures Method is applicable for removing deteriorated concrete cover from reinforced members such as pavements and decks where it is unlikely that the reinforcement will be contacted Boom allows removal from vertical and overhead surfaces Concrete containing wire mesh can be cut without significant losses in productivity Method produces relatively small debris that is easily handled	Removal is limited to concrete outside structural steel reinforcement Significant loss of productivity occurs in sound concrete Productivity is significantly reduced when boom is operated from top of wall as operator's view of cutting is very limited Concrete that remains may be damaged (microcracking) Skid loader units typically mill a more uniform removal profile than other rotary-head and water-jet units Noise, vibration, and dust are produced

Presplitting	Chemical presplitting, expansive agents	Method is applicable for presplitting concrete members where depth of boreholes is 10 times borehole diameter or greater Expansive products can be used to produce vertical presplitting planes of significant depth Some products form a clay-type material when mixed with water that allows the material to be packed into horizontal holes No vibration, noise, or flying rock is produced other than that produced by the drilling of boreholes and the secondary breakage method	Personnel must be restricted from presplitting area during early hours of product hydration as material has the potential to blow out of boreholes and cause injury Presplitting with expansive agents is typically costly Expansive products that are prills or become slurries when water is added are best used in gravity-filled, vertical, or near-vertical holes. A liner may be required to contain the expansive material in holes drilled into concrete with extensive cracks Products are limited to a specific temperature range Rotary-head milling or mechanical-impacting methods will be required to complete removal Development of presplitting plane is significantly decreased by presence of reinforcing steel normal to plane Loss of control of presplitting plane can result if boreholes are too far apart or holes are located in severely deteriorated concrete
	Mechanical presplitting, piston-jack splitter	Method is applicable for presplitting more massive concrete structures where 250 mm (10 in.) or more of face is to be removed and presplitting requires boreholes of a depth greater than can be used by plug-and-feather splitters	Large-diameter (90-mm (3-1/2-in.)) boreholes are required that increase cost Splitters are typically used in pairs to control presplitting plane Hand-held breakers and pry bars are typically required to complete removal

(*Continued*)

Table 5.2 (Continued)

Category	Method	Features	Considerations
		Splitter can be reinserted into boreholes to continue removal for full depth of holes Splitter can be used in areas of difficult access No vibration, noise, or flying rock is produced other than that produced by the drilling of boreholes and the secondary breakage method	Development of presplitting plane is significantly decreased by presence of reinforcing steel normal to presplit plane Loss of control of presplitting plane can result if boreholes are too far apart or holes are located in severely deteriorated concrete Availability of splitters is limited in the U.S.
	Mechanical presplitting, plug-and-feather splitter	Method applicable for presplitting slabs, walls, and other concrete members where presplitting depth is 4 ft or less Method typically less costly than cutting methods Initiation of direction of presplitting can be controlled by orientation of plug and feathers Splitters can be used in areas of limited access Limited skills required by operator No vibration, noise, or flying rock is produced other than that produced by the drilling of boreholes and the secondary breakage method	Splitter cannot be reinserted into boreholes to continue presplitting after presplit section has been removed, as the body of the tool is wider than the borehole Development of presplitting plane in direction of borehole depth is limited Development of presplitting plane is significantly decreased by presence of reinforcing steel normal to plane Secondary means of breakage will typically be required to complete removal Loss of control of presplitting plane can result if boreholes are too far apart or holes are located in severely deteriorated concrete

Courtesy of the United States Army Corps of Engineers

BLASTING

Blasting is done with boreholes that use rapidly expanding gas to create controlled fracture and concrete removal. The use of explosive blasting is usually not considered for jobs that require the removal of less than 10 inches of concrete from the face of a structure.

Blasting is a fast, and often cost-effective, method of removing concrete, but there is a downside. When blasting is used, there is a risk of damaging sound concrete that is not meant to be disturbed. A solution to this is a method known as smooth blasting.

> Whenever there is a possibility of lateral forces causing a transfer of moment on columns, the shear resulting from this action must be prohibited by the use of reinforcement.

Smooth blasting uses detonating cord to distribute blast energy throughout a borehole. This can avoid energy concentrations that might damage surrounding structural elements. An extended version of smooth blasting is known as cushion blasting, which is essentially the same as smooth blasting, with the exception that cushion blasting requires the filling of boreholes with wet sand. Cushion blasting is seldom used.

Cutting is often used in conjunction with blasting. Selected cuts are made at prime locations to control the breakaway of concrete when blasting is employed. Sequential blasting allows for more delays to be used with each firing. This optimizes the amount of explosive detonated with each firing. Air-blast pressures are maintained at acceptable levels, as are ground vibrations and fly rock.

CRUSHING

Crushing is done with hydraulically powered jaws. Boom-mounted mechanical crushers and portable mechanical crushers are both used. When total demolition is desired, a boom-mounted crusher is normally used. This type of equipment is most effective when not used for partial removal. Portable crushers work best when the shearing plane depth is 12 inches or less.

CUTTING

Cutting is used to remove full sections of concrete. The following methods can be used for cutting:

- Abrasive water jets
- Diamond saws
- Diamond-wire
- Stitch drilling
- Thermal tools

Abrasive water jets can be used to make cutouts through slabs, walls, and other concrete members. This type of cutting is capable of cutting through steel reinforcing and hard aggregates. A drawback to abrasive water cutting is that it tends to be slow and costly. Other concerns are controlling the waste water and maintaining personal safety when working with very noisy, high-pressure systems.

Diamond blades are used when the cutting depth is less than 2 feet. Blade selection is dependent on the type of aggregate and reinforcing that must be cut. The time required for blade cutting depends on the hardness of the material being cut, but the process can be slow.

Diamond-wire cutting is used when the cutting depth is greater than a diamond blade can handle. This type of cutting is well suited for cuts where access is limited or difficult. Wire cutting is specialized and tends to be expensive, but sometimes it is the best alternative available.

> Shearhead arms must not be interrupted within a column section.

STITCH CUTTING

Stitch cutting is used when the cut depth is greater than what can be reached with a diamond saw blade. This is most often the case when only one face of the concrete is accessible. If two faces are accessible, wire cutting is more likely to be used.

THERMAL CUTTING

Thermal cutting is used for cutting through heavily reinforced concrete structures where site conditions will allow efficient flow of molten concrete from cuts. Flame tools are favored for this type of cutting for depths up to 2 feet. Cutting lances are used when the cutting depth is greater. Thermal cutting is expensive and is limited in places where it can be used. The dangers associated with thermal cutting include working with compressed and flammable gases, hot flying rock, and high temperatures.

> Negative movement reinforcement in a continuous member must be anchored in or through the supporting member by embedment length, hooks, or mechanical anchorage.

IMPACTING METHODS

Impacting methods involve the use of some sort of mass to hit concrete repeatedly. A simple example of this would be banging concrete with a sledgehammer.

The impact is meant to fracture and spall concrete. Reinforcements must be cut out when impacting methods are used for concrete removal.

BOOM-MOUNTED CONCRETE BREAKERS

Boom-mounted concrete breakers can be found attached to backhoes. This type of breaker is fast and cost-effective. The breaker may be operated by compressed air or hydraulic pressure. Saw cuts are typically used to outline the breakout area. This reduces feathered edges. Concrete that remains after this type of breaking is done can suffer from microcracking, so one must check for that. If it is found, a high-pressure water jet might remove microfractured concrete. The water pressure should be set at a minimum of 20,000 pounds per square inch (psi).

> Continuous reinforcement is required at interior supports of deep flexural members to provide negative movement tension.

FIGURE 5.1

Installing concrete around reinforcing bars.
Courtesy of www.reedpumps.com

SPRING-ACTION HAMMERS

Spring-action hammers are normally used on thin concrete. This method is most appropriate for total demolition. Cutting concrete along a boundary for a cutout is recommended when spring-action hammers are used. Microcracking along with exposed reinforcing steel are likely to occur.

HANDHELD IMPACT BREAKERS

Handheld impact breakers work well for limited concrete removal. These are basically jackhammers. The breaker may be powered by compressed air, hydraulic pressure, self-contained gasoline engines, or self-contained electric motors. Small, shallow sections of concrete can be demolished with great mobility when a handheld breaker is used.

Both mechanical and welded splices are allowed by code.

HYDROMILLING

Hydromilling is sometimes called hydrodemolition or water-jet blasting. This type of concrete removal is effective on concrete with a depth of up to 6 inches. Sound concrete and reinforcements are not normally damaged when hydromilling is used. However, this method is expensive and can be slow when sound concrete is encountered. Another problem with this method is the likelihood of blowouts. A blowout is a hole through a concrete member. A large amount of portable water must be available. Some units use up to 1,000 gallons of water per hour. Flying rock is also a common danger.

ROTARY-HEAD MILLING

Rotary-head milling is used to remove deteriorated concrete from mass structures. If the compressive strength of concrete is 8,000 psi or greater, rotary-head milling is not practical. Remaining concrete can suffer from microcracking.

When feasible, splices should be located away from any points of maximum tensile stress.

PRESPLITTING

Presplitting depends on wedging forces in a designed pattern of boreholes to produce a controlled cracking of concrete to facilitate removal of concrete by other means. The extent of presplitting planes is affected by the pattern, spacing, and depth of boreholes. Chemical-expansive agents and hydraulic splitters are used in presplitting. Reinforcing steel significantly decreases the presplitting plane. A loss of control can occur if boreholes are too far apart or are drilled in severely deteriorated concrete.

CHEMICAL AGENTS

Chemical agents can be used for presplitting concrete. For this to be used, the depth of boreholes should be ten times the diameter of boreholes, or more. This method is expensive. However, it works very well when presplitting vertical planes of significant depth. In the early hours of use, chemical agents can blow out boreholes and cause personal injury. It is not uncommon for rotary-head milling or mechanical-impacting methods to be used in conjunction with chemical presplitting methods to obtain a completed job.

When dealing with spirally reinforced compression members, lap length in a splice must not exceed 12 inches.

PISTON-JACK SPLITTERS

If you are presplitting a concrete face that is 10 inches or thicker, a piston-jack splitter is a good way to go. Boreholes for this type of work must have a minimum diameter of 3.5 inches. The cost of using piston-jack splitters can be prohibitive.

PLUG-FEATHER SPLITTER

What the heck is a plug-feather splitter, you may ask? Plug-feather splitters are used on concrete surfaces that have a depth of no more than 4 feet. The direction of presplitting can be controlled with this type of splitter. The body of this type of splitter is wider than the borehole it is used on. Consequently, the splitter cannot be reinserted into boreholes to continue presplitting after a section has been presplit.

PREP WORK

The prep work done prior to a repair is one of the most important steps in making successful concrete repairs. Preparation work varies. It depends on the type of repair being made. In general, the concrete being prepared should be sound, clean, rough-textured, and dry. There are exceptions, but this is the normal rule of thumb.

FIGURE 5.2

Working with a prepared site.
Courtesy of www.reedpumps.com

Normally, all deteriorated concrete should be removed prior to a repair. This is often done with impact tools, either handheld or boom-mounted. After secondary removal, sandblasting, wet or dry, or water-jet blasting can be used to clean the sound concrete. These are the options for cleaning concrete for a repair:

- Chemical cleaning
- Mechanical cleaning
- Shot blasting
- Blast cleaning
- Acid etching
- Bonding agents

Chemical cleaning

Chemical cleaning is needed when concrete is contaminated with oil, grease, or dirt. Detergents and other concrete cleaners can be used to rid the concrete of contaminants. However, the cleaners themselves must also be removed before a repair is made. Avoid the use of solvents because they may push contaminants deeper into the concrete. While muriatic acid is used for etching concrete, it is not very effective for removing grease or oil.

A *middle strip* is a strip that it is bounded by two column strips.

Mechanical cleaners

Mechanical cleaners include scabblers, scarifiers, and impact tools. Different heads on the tools allow for different types of abrasive material. In any event, secondary cleaning will be necessary after using mechanical cleaners. This is done with sandblasting, wet or dry, or water-jetting.

Shot blasting

Thin overlay repairs cry out for shot blasting. This is the use of steel shot being blasted against the concrete to create a uniform surface. Once the blasting is done, the steel shot is gathered by a vacuum and saved for later use. The result is a dry surface that is ready to accept a repair.

FIGURE 5.3

Pressurized spraying.
Courtesy of www.reedpumps.com

Blast cleaning

Blast cleaning is done with water-jetting and both wet and dry sandblasting. Sandblasting requires the use of an effective oil trap to prevent contamination of concrete surfaces during the cleaning operation. Water-jetting equipment with operating pressures of 6,000 to 10,000 psi is commercially available for cleaning concrete.

Acid etching

Acid etching is used to remove laitance and normal amounts of dirt. Cement paste is removed by the acid. This provides a rough surface. General opinion is that acid etching should be used only when other choices are not suitable.

> Every middle strip must be designed and installed to proportion the resistance of the sum of movements assigned to its two half-middle strips.

Bonding agents

Bonding agents should be used on repairs that are less than 2 inches thick. Thicker repairs can be made without bonding agents when the receiving surface it properly prepared. Many types of bonding agents are available. Always refer to the manufacturer's recommendations for use of their product.

REINFORCING STEEL

Corrosion is the enemy of reinforcing steel. When the steel reinforcing material has to be replaced, concrete must be removed from around the existing reinforcement. A jackhammer is the tool of choice in most such cases. All weak, damaged, and easily

FIGURE 5.4

Working with reinforcing steel.
Courtesy of www.reedpumps.com

removable concrete should be removed. Circumstances may allow existing reinforcing steel to be cleaned and left in place. To do this, at least one-half of the existing steel must not be corroded.

If an air compressor is used to clean corrosion, make sure the compressor is either an oil-free compressor or a compressor that has an oil trap. Otherwise, the concrete may be stained with blowing oil. Dry sandblasting is the best method for cleaning reinforcing steel. Wet sandblasting and water-jetting can also be effective.

When walls or columns are built integrally with a slab system, they must be constructed to resist factored loads on the slab system.

ANCHORS

Anchors require the drilling of holes. This is usually done with a rotary carbide-tipped drill bit or a diamond-studded drill bit. Using a jackhammer to create anchor holes is not recommended because of potential damage to in-place concrete. Anchor holes should be cleaned out and protected from debris entering the holes.

Bonded anchors can be headed or headless bolts, threaded rods, or deformed reinforcing bars. These anchors may be either grouted anchors or chemical anchors. Grouted anchors are embedded in predrilled holes with neat Portland cement, Portland cement and sand, or other commercially available premixed grout. An expansive grout additive and accelerator are commonly used with cementitious grouts.

FIGURE 5.5

Building a reinforced wall.
Courtesy of www.reedpumps.com

Chemical anchors are embedded in predrilled holes with two-component polyesters, vinylesters, or epoxies. These anchors come in glass capsules, plastic cartridges, tubes, and bulk. Once inserted into a predrilled hole, the chemical anchor casing is broken or opened. The two components mix. Epoxy products are used in bulk systems. The epoxy is mixed in a pot or pumped through a mixer and injected into a hole.

Expansion anchors are placed in predrilled holes and expanded by tightening a nut, hammering the anchor, or expanding into an undercut in the concrete. Expansion anchors that rely on side-point contact to create frictional resistance should not be used where anchors are subjected to vibratory loads. Certain wedge-type anchors don't perform well when subjected to impact loads. Undercut anchors are suitable for dynamic and impact loads.

After any anchors are placed and ready for use, the anchor's strength should be tested. This is a step that should not be overlooked. It is desirable to specify a maximum displacement in addition to the minimum load capacity. This completes our work here.

CHAPTER

Materials and methods for repair and rehabilitation

6

Some types of concrete failures can be repaired by adding additional reinforcement. For example, a bridge girder that is failing may be saved by posttensioning. Cracks might be able to be overcome with drillholes at 90-degree angles to the cracks. There are many ways to work with damaged concrete. Our goal in this chapter is to review both the materials and methods you can choose from when repairing and rehabilitating concrete. I just mentioned the use of drillholes at 90-degree angles to strengthen cracked concrete. To give you an example of what to expect in this chapter, let's explore the procedure more closely. The goal is to seal any cracks and add reinforcement bars. Holes are drilled with a three-quarter-inch diameter at 90-degree angles to the crack plane. Holes must be cleaned of all loose dust. The drilled holes and the crack plane are filled with an adhesive that is usually an epoxy. This is done by pumping the adhesive into the holes and crack plane with low pressure that typically ranges from 50 to 80 psi. Reinforcing bars, using either Number 4 or Number 5 bars, are placed in the drilled holes. The rods should extend at least 18 inches on each side of the crack. When the components are all in place, the adhesive bonds the bar to the walls of the hole, fills the crack plane, and bonds the cracked concrete surfaces together in one monolithic form. This reinforces the compromised concrete section.

Temporary elastic crack sealant is required for a successful repair. Gel-type epoxy crack sealants work well within their elastic limits. When cold weather is a factor, silicone or elastomeric sealants work well. Sealant should be applied in a uniform layer approximately 1/16 inch to 3/32 inch. This layer should span the crack by a minimum of three-quarters of an inch on each side.

Reinforcing bar placement is subject to individual repair requirements. The spacing and pattern of the bars will be determined by a repair design that is relevant to the fault at hand. Bars can be used in a number of ways to reinforce damaged concrete. Let's look at another example.

> All concrete walls must be designed to accept eccentric loads and any lateral or other loads that may have an impact on them.

doi: 10.1016/B978-1-85617-549-4.00006-X

Reinforcement bars can be used externally, and in conjunction with other means, to place new concrete as a reinforcement. Longitudinal reinforcing bars and stirrups or ties around concrete members can be encased with shotcrete or cast-in-place concrete. Girders and slabs have been reinforced by addition of external tendons, rods, or bolts that are prestressed.

FIGURE 6.1

Concrete building components.

Copyright © Gary S. Figallo. Courtesy of Faddis Conrete Products

PRESTRESSING STEEL

You can create compressive force by using prestressing steel strands or bars. When this is done, there must be adequate anchorage and you have to be sure that the compressive force will not adversely affect other portions of a concrete member.

Cracks found in slabs on grade can sometimes be repaired with steel plates. To do this, concrete is cut across the cracks. The cuts are usually 2 to 3 inches deep, and they tend to span the crack, so there are 6 to 12 inches of the cut on each side of the cracks. The cuts are filled with epoxy, and steel plates are forced into the epoxy-filled cracks. The result is new strength in the cracked concrete.

AUTOGENOUS HEALING

Autogenous healing is a natural process of crack repair that can occur in the presence of moisture and the absence of tensile stress. This type of healing does not occur with active cracks, but it can occur with dormant cracks. While moisture is needed for the process to occur, moving water is not the type of moisture you want. Running water tends to dissolve and wash away lime deposits. A damp moisture that evaporates is suitable for autogenous healing.

The healing that takes place is a result of carbonation of calcium hydroxide in cement paste by carbon dioxide that is present in surrounding air and water. You will find that calcium carbonate and calcium hydroxide crystals precipitate, accumulate, and grow within the cracks of concrete. This is a type of chemical bonding that restores some of the strength of cracked concrete. To develop substantial strength, water saturation of the crack and adjacent concrete is needed. Continuous saturation speeds the healing process. Even a single cycle of drying and reimmersion can result in a drastic reduction in the degree of healing that is accomplished.

FIGURE 6.2

Concrete building components.
Copyright © Gary S. Figallo. Courtesy of Faddis Conrete Products

Generally speaking, the thickness of nonbearing walls must not be less than 4 inches.

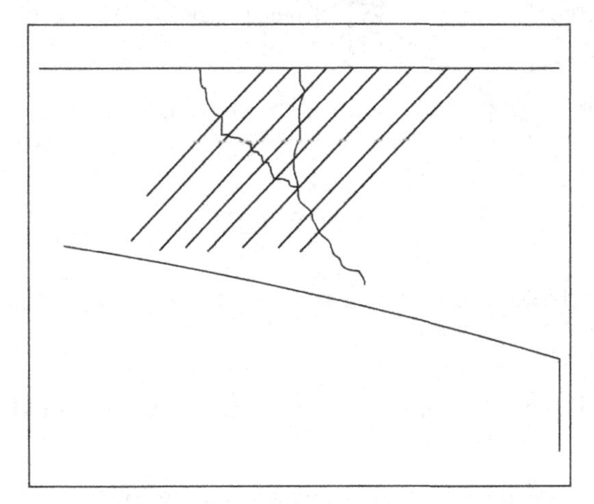

FIGURE 6.3

Crack repair using conventional reinforcement with drill holes 90 degrees to the crack plane.
Courtesy of United States Army Corps of Engineers

CONVENTIONAL PLACEMENT

Conventional placement of concrete is a method of using new concrete to repair damaged concrete. The repair concrete must be able to make an integral bond with the base concrete. A low w/c and a high percentage of coarse aggregate is needed in the repair concrete to minimize shrinkage cracking.

Concrete replacement should be used when defects extend through a wall or beyond the reinforcement structure within the concrete. It is also a desirable solution when there are large sections of honeycombing in concrete. Replacement concrete should not be used when there is an active threat of deterioration that caused the existing concrete to fail.

Concrete repair with new concrete requires the removal of existing, damaged concrete. The goal is to get down to solid concrete that the new concrete can bond with. Generally, the depth desired is about 6 inches. A light hammer is normally used as sound concrete is found. This is done to clean the surface of the good concrete.

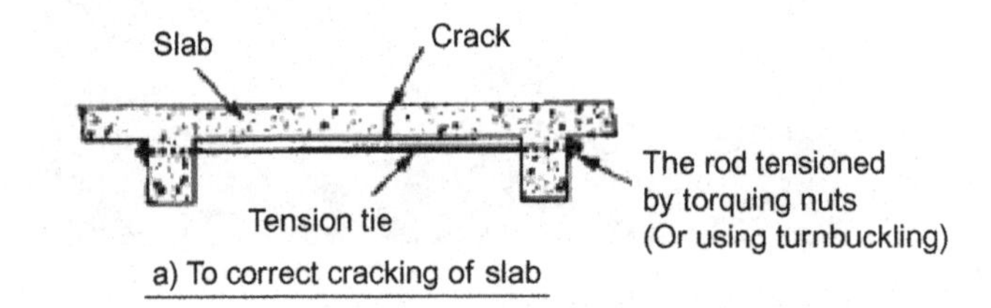

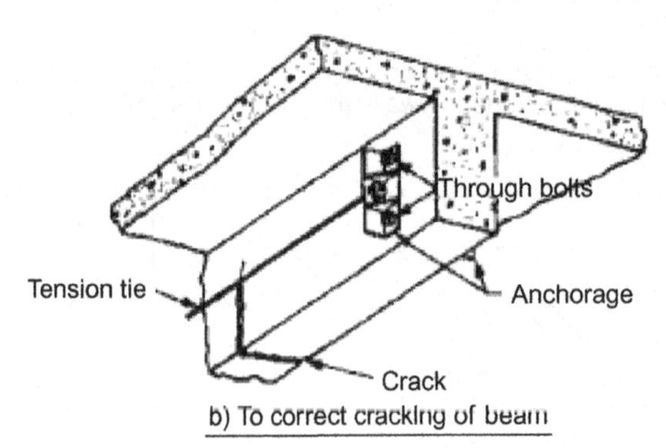

FIGURE 6.4

Crack repair with use of external prestressing strands or bars to apply a compressive force.
Courtesy of United States Army Corps of Engineers

When repairing vertical sections of concrete, the cavity should have the following specifications:

- A minimum of spalling or featheredging at the periphery of the repair area
- Vertical sides and horizontal top at the surface of the member

- Inside faces generally normal to the formed surface, except that the top should slope up toward the front at about a 1:3 slope
- Keying as necessary to lock the repair into the structure
- Sufficient depth to reach at least 1/4 inch, plus the dimension of the maximum size aggregate behind any reinforcement
- All interior corners rounded with a radius of about 1 inch

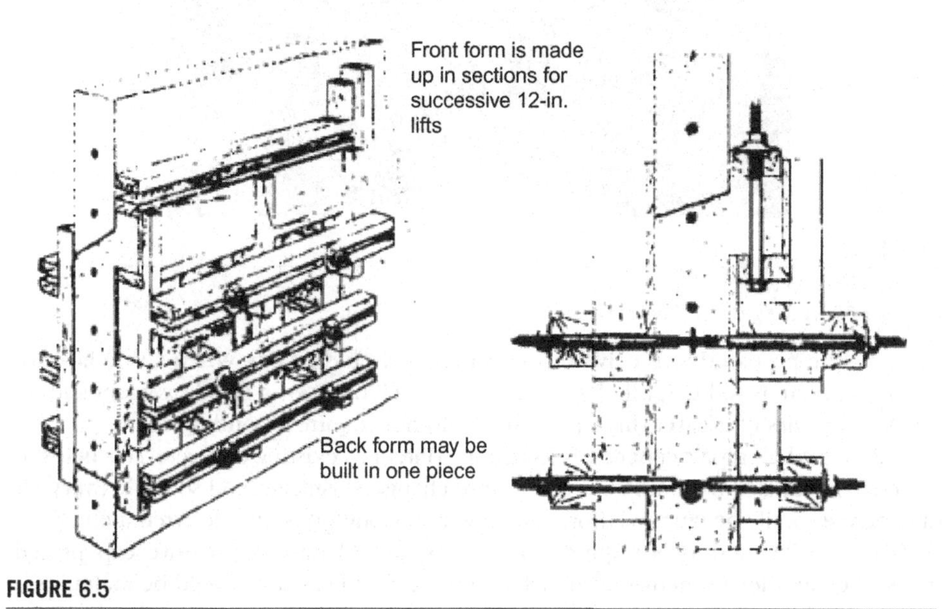

FIGURE 6.5

Detail of form for concrete replacement in walls after removal of all unsound concrete.
Courtesy of United States Army Corps of Engineers

All sound concrete must be clean before repair concrete is applied. This is best accomplished with sandblasting, shotblasting, or an equally acceptable process. Only the surface that is to receive new concrete should be sandblasted. Final cleaning should be done with compressed air or water. It is common for dowels and other reinforcements to be installed to make a concrete patch self-sustaining and to anchor it to the underlying concrete to provide an added safety factor.

> When calculating the moments and shears permitted on footings on piles, you can assume that the reaction from any pile will be concentrated on the center of the pile.

When repairing large vertical sections with new concrete, forming will be necessary. The forms must be strong and mortar-tight. Front panels of a form should be constructed as placing progresses so the concrete can be conveniently placed in lifts. If a back panel is needed for a form, it can be a single section.

FIGURE 6.6

Vertical concrete walls.

Copyright © Gary S. Figallo. Courtesy of Faddis Concrete Products

Concrete prepared to receive repair concrete should be dry. When a thin layer of repair concrete is to be applied—say, 2 inches thick or less—a bonding agent should be used. Repairs of greater thickness usually do not require a bonding agent.

It is best when repair concrete is similar in content to existing concrete. This helps to avoid strains caused by temperature, moisture change, shrinkage, and so forth. Every lift of concrete should be vibrated thoroughly. Internal vibration is the preferred method.

If external vibration is required, the cavity should have a pressure cap placed inside the chimney immediately after filling the cavity. Pressure should be maintained

FIGURE 6.7

Precast concrete wall.

Copyright © Gary S. Figallo. Courtesy of Faddis Concrete Products

during the vibration. This type of vibration should be repeated at 30-minute intervals until the concrete hardens and no longer responds to vibration. The projection left by the chimney is normally removed on the second day after the pour. And, of course, proper curing is essential.

FIGURE 6.8

Rigging precast concrete members for placement.
Copyright © Gary S. Figallo. Courtesy of Faddis Concrete Products

> Precast members must be designed to withstand the forces and deformations that occur in and adjacent to connections.

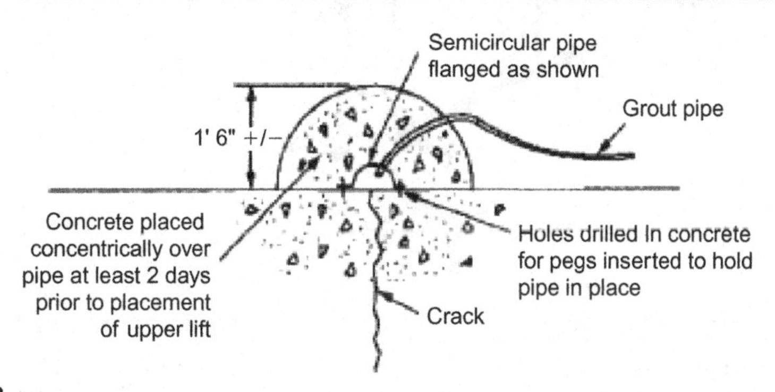

FIGURE 6.9

The use of a semicircular pipe in the crack arrest method of concrete repair.
Courtesy of United States Army Corps of Engineers

CRACK ARREST TECHNIQUES

Crack arrest techniques are used to stop cracking that can be caused by restrained volume change of concrete installations. The techniques are not suitable for cracks created by excessive loading. These techniques are typically used during the construction of massive concrete structures.

One simple technique that can be used consists of installing a grid of reinforcing steel over a cracked area. The reinforcing steel is then surrounded by conventional concrete rather than the mass concrete being used in the structure. Another, somewhat more complex, method involves the use of a piece of semicircular pipe. The pipe starts out with an 8-inch diameter and is cut in half. It is a 16-gauge pipe that is bent into a semicircular shape with about a 3-inch flange on each side. The area surrounding cracked concrete should be clean and the pipe section should be centered on the crack. Sections of the pipe are then welded together. Holes are cut into the pipe to receive grout pipes. Then the pipe section is covered with concrete placed concentrically by hand methods. Grout pipes can be used for grouting at a later date to attempt to restore structural integrity of the cracked section.

When spacing transverse ties perpendicular to floor or roof slabs, the spacing must not exceed the distance of the spacing of the bearing walls.

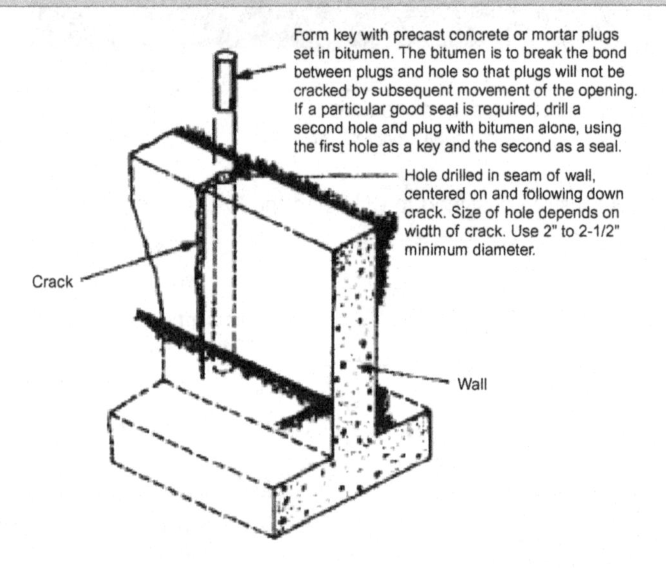
Form key with precast concrete or mortar plugs set in bitumen. The bitumen is to break the bond between plugs and hole so that plugs will not be cracked by subsequent movement of the opening. If a particular good seal is required, drill a second hole and plug with bitumen alone, using the first hole as a key and the second as a seal.
Hole drilled in seam of wall, centered on and following down crack. Size of hole depends on width of crack. Use 2" to 2-1/2" minimum diameter.
Crack
Wall

FIGURE 6.10

Repair of crack by drilling and plugging.
Courtesy of United States Army Corps of Engineers

DRILLING AND PLUGGING

Drilling and plugging a crack involve drilling down the length of the crack and grouting it to form a key. This procedure is normally used on cracks that run basically in a straight line and that are accessible at one end. Vertical cracks in walls are the most likely cracks to be corrected with drilling and plugging.

When drilling and plugging, a hole with a diameter of 2 to 3 inches is drilled into the center of a crack, following the direction of the crack.The diameter has to be large enough to intersect the crack along its full length and to provide enough repair material to structurally take the loads exerted on the key. Drilled holes have to be cleaned and then filled with grout. Once the grout key is in place, it prevents transverse movement of sections of concrete that are adjacent to a crack. Another function of the grout key is to reduce heavy leakage through a crack and to reduce the loss of soil from behind a leaking wall.

On occasions when watertightness is essential and structural load transfer is not, the drilled hole should be filled with a resilient material of low modulus, such as asphalt or polyurethane form in lieu of Portland cement grout. If you must deal with watertightness and the keying effect, you can use resilient material in a second hole and grout the first hole.

DRYPACKING

When you ram or tamp a low-w/c mortar into a confined area, you are drypacking. There is minimal shrinkage with this type of repair. Patches are normally tight and of good quality. By good quality, I mean that they are normally durable, strong, and watertight. No special equipment is required for drypacking, and this makes it a preferred method when it can be used.

Circumstances that call for drypacking include patching rock pockets, form tie holes, and small holes with a relatively high ratio of depth to area. Shallow defects are not suitable for drypacking, and active cracks should not be treated with drypacking.

> A nominal strength in tension of not less than 16,000 pounds is required when ties are around the perimeter of a floor or roof.

To perform drypacking, first undercut the concrete to be repaired so the base width is slightly greater than the surface width. Dormant cracks should be expanded

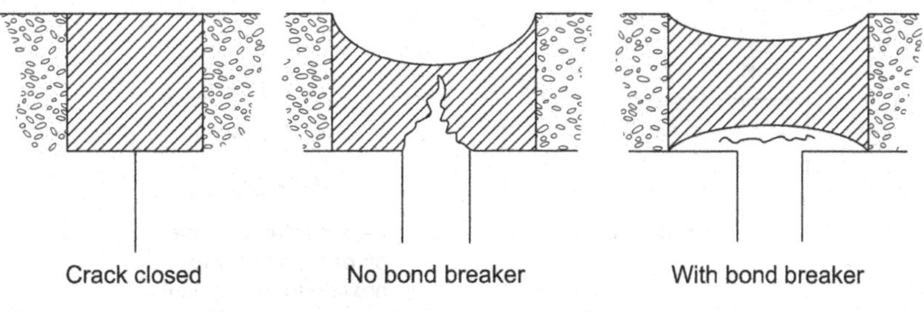

FIGURE 6.11

Effect of bond breaker involving a field-molded flexible sealant.
Courtesy of United States Army Corps of Engineers

to a point where the surface area is about 1 inch wide and 1 inch deep.A power-driven sawtooth bit is the best tool for this procedure. Undercut the slot slightly. Clean and dry the affected area, apply a bond coat, and apply the drypack mortar immediately.

Drypack mortar usually consists of one part cement and two and a half parts sand. (Sometimes there may be three parts sand.) The sand should be capable of passing through a Number 16 sieve. Use only enough water until the mortar sticks together when you squeeze it into a ball by slight pressure with your hands. The mortar ball should leave your hands dry.

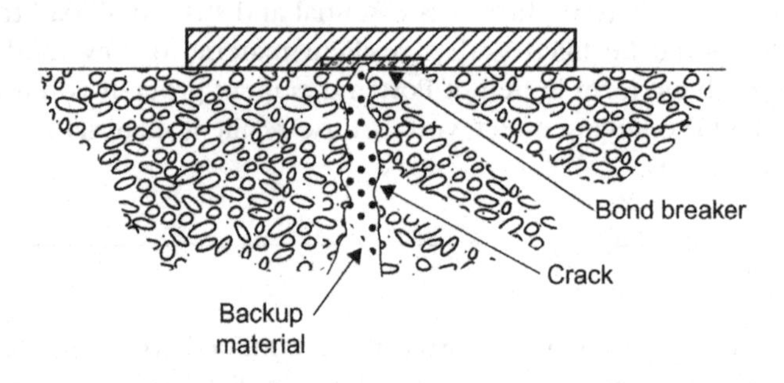

FIGURE 6.12

Repair of a narrow crack with flexible surface seal.
Courtesy of United States Army Corps of Engineers

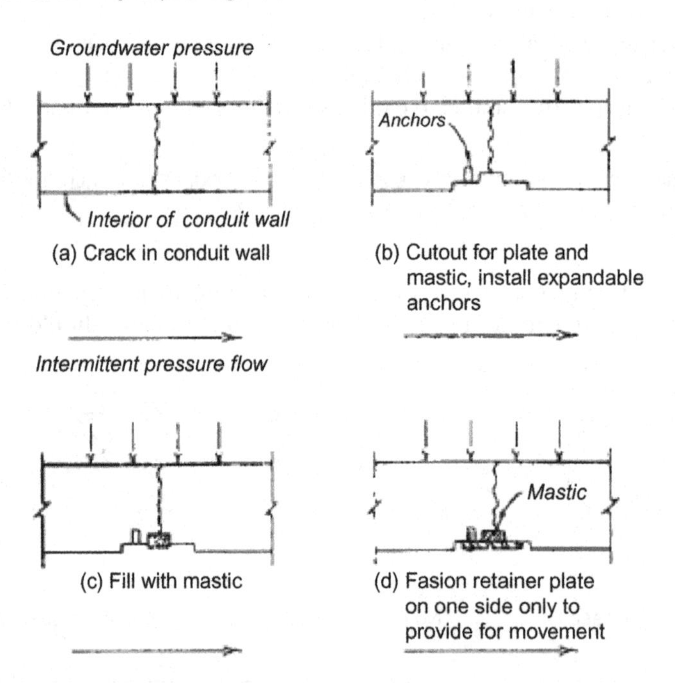

FIGURE 6.13

Repair of crack by use of a retainer plate to hold mastic in place against external pressure.
Courtesy of United States Army Corps of Engineers

Latex-modified mortar and preshrunk mortar are also available. The preshrunk mortar is a low-w/c mortar that has been mixed and allowed to stand idle for 30 to 90 minutes, depending on the air temperature. Remixing is required after the waiting period has expired.

When drypacking is used, the layers should be about $^3/_8$ inch in thickness and should be tamped into place. Portland cement can be used as a part of the mixture if you want to match the color of the existing concrete.

FIBER-REINFORCED CONCRETE

Portland cement that contains discontinuous discrete fibers is known as fiber-reinforced concrete. Fibers are added to the concrete in the mixer. These fibers can be made of steel, plastic, glass, or other natural materials. Fiber-reinforced concrete is often used for the repair of pavement. It is not unusual for increased vibration to be needed for this type of concrete. The prep work for fiber-reinforced concrete is essentially the same as it would be for regular concrete. It can be pumped with up to 1.5 percent fibers by volume in it.

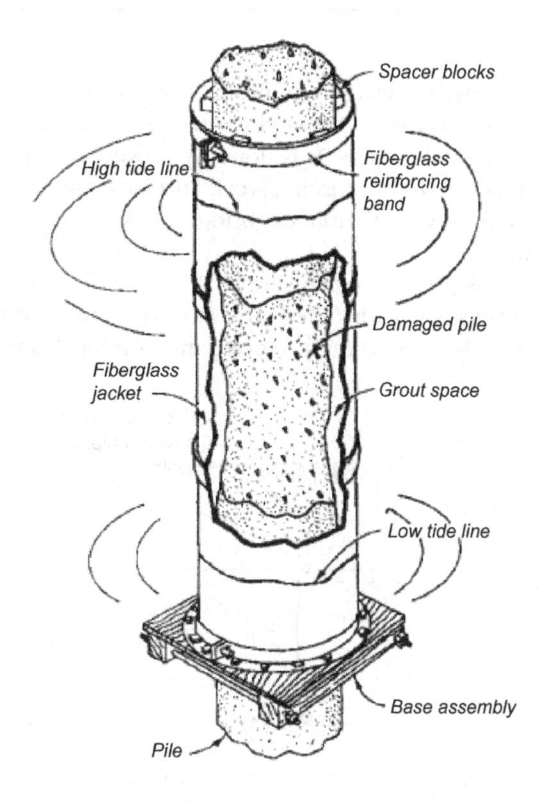

FIGURE 6.14

Typical preformed fiberglass jacket being used in repair of a concrete pile.
Courtesy of United States Army Corps of Engineers

All precast members must be designed and built to withstand curing, stripping, storage, transportation, and erection.

FLEXIBLE SEALING

Flexible sealing involves routing and cleaning a crack and filling it with a suitable field-molded flexible sealant. This is not the same as routing and sealing because an actual joint is made. It is not just a crack being filled. This method can be used to fill active cracks, but don't be fooled. The process is not likely to increase the structural capacity of a cracked section of concrete.

Bond breakers are used at the bottom of a crack slot to prevent a concentration of stress on the bottom of a structure. This can be a polyethylene strip, a pressure-sensitive tape, or some other material that will not bond to the sealant before or during the curing process.

GRAVITY SOAK

High-molecular-weight methacrylate is poured or sprayed onto a horizontal concrete surface and spread by either a broom or squeegee to make a gravity soak. This method penetrates very small cracks. It is done by gravity and capillary action. The purpose is to prevent access to reinforcing steel. Surfaces where gravity soak can be used include horizontal concrete surfaces, bridge decks, parking decks, industrial floors, and pavement.

If you plan to use this process, make sure that the concrete surface is cured and air-dried. It may be necessary with older concrete to clean it of oil, grease, tar, or other contaminants. Sandblasting is the preferred method for this type of cleaning.

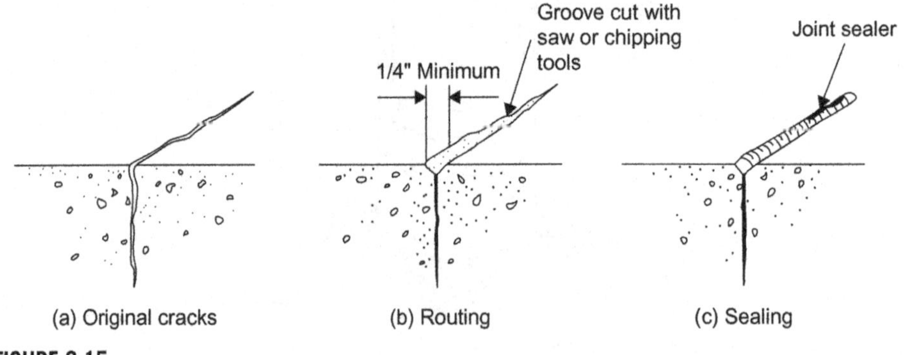

FIGURE 6.15

Repair of crack by routing and sealing.
Courtesy of United States Army Corps of Engineers

Entire composite members, or portions thereof, that are used to resist shear are allowed.

CHEMICAL GROUTING

Chemical grouting consists of two or more chemicals. These chemicals combine to create a gel or solid precipitate, as opposed to cement grouts that consist of suspensions of solid particles in a fluid. This process decreases fluidity and has a tendency to solidify and fill voids in the material into which the grouting is injected. Grouting has pros and cons. Here are the pros:

- They can be used in moist environments.
- There are wide limits of control of gel time.
- It can be used to fill very fine cracks.

And here are the cons:

- A high degree of skill is needed to perform professional grouting.
- There can be a lack of strength in the existing concrete.
- Some grouts dry out too early.
- Some grouts are highly inflammable and cannot be used in enclosed spaces.

HYDRAULIC-CEMENT GROUTING

Hydraulic-cement grouting is a common method of repairing cracks in concrete. It is most often used in dormant cracks and is usually less expensive than chemical grouts. This type of grouting is well suited for large-volume applications. However, the grouting is subject to pressure and may not fill a crack entirely. This type of grouting is normally used for sealing cracks in foundations.

To use hydraulic-cement grouting, you should clean the cracks that are to be sealed. Install built-up seats at intervals across the crack that will provide a pressure-tight contact with the injection tool. Seal the crack between the seats.

HIGH-STRENGTH CONCRETE

A concrete with a 28-day design compressive strength over 6,000 psi is considered a high-strength concrete. Admixtures can be used with the concrete. This type of concrete is used to resist chemical attacks, less abrasion, and improved resistance to freezing and thawing. It is an expensive concrete, but it is worth it in various circumstances. The curing is more critical with high-strength concrete than it is with normal concrete. Water curing is a preferred method with this type of concrete.

Single bars, wire, multiple leg stirrups, and vertical legs of welded wire reinforcement can all be used for ties to transfer horizontal shear.

JACKETING

Jacketing is a method of encasing an existing structural member in new concrete. The member that is being encased can be made of concrete, steel, or wood. Pilings are one of the most common types of structures to benefit from jacketing. This type of repair is especially helpful when a construction member is submerged in water. Jacketing a piling in water can provide additional strength and protection. A form is required for jacketing, and pretty much any type of concrete form can work. Typical forms for jacketing are made from steel, fiberglass, or fabric.

JUDICIOUS NEGLECT

Judicious neglect is a decision to not take any action in the repair of a concrete member. A complete investigation is required before a decision to ignore a problem can be made. Sometimes no action is the best action. This may seem a bit strange, but it is a fact. If existing damage is not likely to escalate and is not causing structural damage, the defect may be monitored as opposed to being repaired.

POLYMER OVERLAYS

The type of polymer overlay used in a repair is generally decided based on the thickness of the overlay. Thin overlays, those about 1 inch thick, are normally done with epoxy mortar or concrete. Thicker overlays, those from 1 inch to 2 inches thick, use a latex-modified concrete. Conventional Portland cement is used when an overlay will be more than 2 inches thick. Epoxy mortars work well for repairs when existing concrete is being attacked by acid or some other aggressive substance.

It is not advisable to use an overlay with a vapor barrier on slab-on-grade applications or concrete walls that are backfilled in freezing climates. Bridge decks are often resurfaced with latex-modified concrete overlays. Epoxy-modified concrete is also used for this type of repair.

The aggregates that are used for conventional concrete can normally be used with epoxy-resin mixtures. All aggregates should be clean and dry when repairs are made. Maximum size for aggregates used in an epoxy-based concrete is 1 inch in diameter. The proportion of aggregate to the mixed resin in epoxy concrete can go as high as a 12-to-1 ratio.

The mixing of epoxy-modified concrete can be done by hand or machine. Large commercial mixers have proved to be very effective in mixing this type of concrete.

When mixing, fine aggregates should be introduced into the mix first and then be followed by coarse aggregates.

> You must neglect the tensile strength of concrete when computing the requirements for reinforcement.

A primer coat of epoxy should be applied to the surface being repaired.This can be done with a brush or a trowel or some other viable method. When the primer is tacky to the touch, it is time to pour the new concrete. Overlays deeper than 2 inches should be made in layers to minimize heat dissipation. Applications made in layers of 2 inches at a time are preferred. A time delay between layers is needed to allow for heat removal. However, epoxy-modified concrete does require prompt action. It doesn't last long in a mixer. A hand tamper can be used to compress layers and speed the application process.

Latex

Latex-modified overlays are often made with styrene-butadiene. This type of admixture is dealt with in much the same manner as conventional concrete is. Temperature, however, can be a problem in hot climates. The latex-concrete should be maintained at a temperature range of between 45°F and 85°F. When air temperatures are higher than 85°F, a night placement of the concrete may be needed to maintain a working temperature range. Another risk of hot weather is rapid drying that results in shrinkage cracks.

A bond coat is typically made by eliminating coarse aggregates from the mix. Then the mixture is broomed onto the concrete surface. Placing of the concrete is done with standard methods. Finish work can be done with vibratory or oscillating screeds, but a rotating cylindrical drum is preferred.

A moistened material is used to cover new concrete. This covering is left in place for one or two days. Then air drying is allowed for about 72 hours. By following these guidelines, you reduce the risk of shrinkage cracks.

> Unbonded tendons require sheathing that is watertight and continuous.

Portland cement

Portland-cement overlays tend to range in depth from 4 to 24 inches. However, thinner layers can be used. Bridge decks often benefit from a concrete overlay. The overlay repairs spalling and surface cracks. Another type of repair where this type of overlay is appropriate is for the repair of concrete that is damaged by abrasion-erosion and deteriorated pavements.

If concrete is damaged by acid or some other aggressive means, Portland cement should not be used for repairs. Active cracking is another defect that should not be repaired with conventional concrete.

POLYMER COATINGS

Polymer coatings aid in the protection of concrete from abrasion, chemical attacks, and freezing-thawing damage. Epoxy is favored for its qualities that allow it to be impermeable to water and to resist chemical attacks. There are types of epoxy resins that can adhere to damp surfaces and sometimes immersed in water.

The preferred temperature range for applying epoxy resins runs between 60°F and 89°F. If you are working outside this range, special precautions are needed. When foot traffic is expected on a finished surface, installers should broadcast a sharp sand to the fresh surface of new concrete.

POLYMER CONCRETE

Polymer concrete (PC) is a composite material. Aggregates are bound together in a dense matrix with a polymer binder. This type of concrete sets up quickly. It bonds well and has strong chemical resistance. Additionally, polymer concrete is high tensile, flexural, and compressive in its strength.

When a polymer is added to Portland cement, the result can be a stronger, more adhesive mixture. Other benefits of mixing polymers with conventional cement include the following:

- Resistance to freezing and thawing
- High degree of permeability
- Improved resistance to chemical attacks
- Better resistance against abrasion and impact

POLYMER IMPREGNATION

Polymer impregnated concrete (PIC) is a Portland cement concrete that is subsequently polymerized. A technique known as a monomer system is used to create this concrete. Water does not mix well with this type of concrete. There is a varying degree of volatility, toxicity, and flammability related to these chemicals. When this admixture is heated, it becomes a tough, strong, durable plastic. When placed in concrete, it adds considerably to a number of qualities. The procedure is very good for filling cracks.

Minnesota Department of Transportation

Concrete Aggregate Worksheet

S.P.	Plant:	Date:	Agg. Source(s) #: FA - CA -
Engineer:	Tester:	Time:	CA - CA -

Sieve Analysis of Coarse Aggregate

Agg. Fract.	CA - ____ Mix Prop. ____ %				CA - ____ Mix Prop. ____ %				CA - ____ Mix Prop. ____ %			
	Test No.		Quality sample submitted.		Test No.		Quality sample submitted.		Test No.		Quality sample submitted.	
	Sample Wt.		By ____ Date		Sample Wt.		By ____ Date		Sample Wt.		By ____ Date	
Sieve sizes Pass - Ret.	Weights		% Pass	Grad. Req.	Weights		% Pass	Grad. Req.	Weights		% Pass	Grad. Req.
	Ind.	Cum.			Ind.	Cum.			Ind.	Cum.		
2"–1 1/2"												
1 1/2"–1 1/4"												
1 1/4"–1"												
1"–3/4"												
3/4"–5/8"												
5/8"–1/2"												
1/2"–3/8"												
3/8"–#4												
#4–Btm												
Check Total		± 0.3% or 0.2 lb of Sample Wt.				± 0.3% or 0.2 lb of Sample Wt.				± 0.3% or 0.2 lb of Sample Wt.		

Coarse Aggregate Percent Passing #200 Sieve Test

(A) Dry weight of original sample	(CA -)		(CA -)		(CA -)
(B) Dry weight of washed sample					
(C) Loss by washing (A – B)					
(D) % Passing #200 (C ÷ A) × 100					

Composite Gradation for (CA -)

Agg. Fract.	CA -	CA -	CA -	Composite	Grad. Req.
Proportions	____ %	____ %	____ %	100%	
2"					
1 1/2"					
1 1/4"					
1"					
3/4"					
3/8"					
#4					

Washing Data for Sieve Analysis of Fine Aggregate

(A) Dry sample and record weight	
(B) Wash and dry sample, record weight	
(C) Loss by washing (A – B)	
Enter (C) to the right, for fine sieve analysis	

* #6 and #200 not included in Fineness Modulus

** #6 is recommended as filler sieve

Sieve Analysis of Fine Aggregate

Quality Sample Submitted. By: ______ Date :				
Test No. ______ Sample Wt.				
Sieve Size Pass Ret.	Weights		% Pass	Grad. Req.
	Ind.	Cum.		
3/8"–#4				100
#4–#6				95–100
*#6–#8				**
#8–#16				80–100
#16–#30				55–85
#30–#50				30–60
#50–#100				5–30
#100–#200				0–10
*#200–Btm				0–2.5
Loss by washing				
Check Total		± 0.3% of Sample Wt.		
Fineness Modulus	Within ± 0.20			

FIGURE 6.16

Concrete aggregate worksheet.

Courtesy of Minnesota Department of Transportation

> Any posttensioning ducts installed in areas subject to freezing temperatures must be protected against water accumulation in the ducts.

POLYMER INJECTION

Polymer injections can be rigid or flexible systems. Epoxies are a common type of rigid system. These are often used for repairing cracks. A polyurethane system is much more flexible. This type of system is used on active cracks and to stop water infiltration. A high-pressure injection is preferred.

Rigid repairs have been used for cracks in bridges, in buildings, dams, and other types of concrete structures. Normally, a rigid repair is made only on dormant cracks. There are some exceptions, but a rigid repair is not normally recommended for active cracks. This type of repair can be used for delaminations in bridge decks.

Active cracks should be filled with flexible repairs. The grout used will make the crack a joint that can move. There are water-activated polyurethane grouts. They can be hydrophobic or hydrophilic. High-pressure injections are typically done at pressures of 50 psi, or higher. These are the steps for an injection:

1. Clean cracks.
2. Seal surfaces.
3. Drill holes to insert injection fittings.

FIGURE 6.17

Precast concrete.

Copyright © Gary S. Figallo. Courtesy of Fassis Concrete Products

4. Bond cracks that are not V-grooved with a flush fitting.
5. Mix the filler.
6. Inject the filler.
7. Remove any surface seals.

PRECAST CONCRETE

Precast concrete has several benefits in the repair of damaged concrete structures: ease of construction, rapid construction, high quality, durability, and economy. Precast concrete is traditionally used for the following types of repairs:

- Navigation locks
- Dams
- Channels
- Flood walls
- Levees
- Coastal structures
- Marine structures
- Bridges
- Culverts
- Tunnels
- Retaining walls
- Noise barriers
- Highway pavement

FIGURE 6.18

Precast noise barrier.

FIGURE 6.19

Concrete deck.

PREPLACED-AGGREGATE CONCRETE

Preplaced-aggregate concrete is used to fill in voids in concrete. It is injected with a Portland cement–sand grout and can be pumped into a form to fill voids. Water in the voids is displaced, and the concrete forms a solid mass. This type of repair is normally used on large repairs. There is a slow shrinkage rate when this method is used.

RAPID-HARDENING CEMENT

Rapid-hardening cement has a minimum compressive strength of 3,000 psi that is developed in approximately 8 hours or less. Some cements can obtain strength in as little as 1 hour. Magnesium-phosphate cement is one such cement. You have to work fast with this product, but it reduces downtime on a job site.

High alumina cement can have as much strength in 24 hours as conventional concrete will have after 28 days of setting. High humidity and air temperatures of 68°F or greater can reduce the strength of this mixture. Some other types of fast-setting cements that are available are regulated-set Portland cement, gypsum cement, special blended cements, and packaged patching materials.

ROLLER-COMPACTED CONCRETE

Roller-compacted concrete is concrete that is compacted by a roller before it is completely hardened. This type of concrete is best suited for large placement areas where there is little to no reinforcement or embedded metals. The new construction of dams and pavement is a suitable application for roller-compacted concrete. It is also used to repair dams. Its benefits are added strength and stability.

ROUTING AND SEALING

Routing and sealing is a common method of repairing dormant cracks. The procedure should not be used on active cracks. A minimum surface width for a crack to be routed and sealed is one-quarter inch. When you are dealing with pattern cracks or narrow cracks, the routing will enlarge the cracks to make them suitable for sealants. Sealants are used to prevent water infiltration.

SHOTCRETE

Shotcrete is mortar that is pneumatically projected at high velocity onto a surface. It can contain coarse aggregate, fibers, and admixtures. When shotcrete is used, the result can be an excellent bond and a viable repair. It can be used for bridges, buildings, lock walls, dams, and hydraulic structures.

SHRINKAGE-COMPENSATING CONCRETE

Shrinkage-compensating concrete is an expansive cement concrete that is used to repair cracks. The concrete expands in volume after setting and during hardening. With the proper reinforcement and containment, good strength can be obtained.

We are now ready to move on to the topic of maintenance.

CHAPTER 7

Maintenance of concrete

The maintenance of concrete is the answer to reducing expenses associated with concrete repairs. Prevention is the best medicine. As obvious as this should be, it is too often ignored. Routine maintenance can prevent, or at least postpone, the need for costly and time-consuming rehabilitation and repairs. Typical maintenance can be expensive, but it is far less expensive than major repairs. Money invested in advance may not be a desirable option, but it is often better than the alternative.

STAINS

Stains on concrete can be a sign of trouble. They can penetrate the concrete if it is porous and absorbent. For this reason, not to mention eye appeal, stains should be removed. However, there are many different types of stains, some of which include the following:

- Iron rust
- Oil
- Grease
- Dirt
- Mildew
- Asphalt
- Efflorescence
- Soot
- Graffiti

Not all stains can be treated equally. Removal methods vary. Choosing a cleaning method depends on what type a stain is being dealt with. It's best if you know what the stain is, but when identification is not possible, you can test different types of cleaning options. The cleaning components can be tested on small parts of the stained concrete. Try testing the following substances in this order: organic solvents, oxidizing bleaches, reducing bleaches, and finally acids.

doi: 10.1016/B978-1-85617-549-4.00007-1

Table 7.1 Concrete Coatings that Prevent Chemical Attack and Reduce Moisture Penetration (NACE 1991)

Coating	Water Repellancy	Cleanability	Aesthetic	Concrete Dusting	Mild Chemical	Severe Chemical	Moderate Physical	Severe Physical
Silicones/Silanes/ Siloxane	R	NR	NR	NR	NR	NR	NR	NR
Cementitious	R	NR	R	NR	NR	NR	NR	NR
Thin-Film Polyurethane	R	R	R	R	R	NR	R	NR
Epoxy Polyester	R	R	R	R	R	NR	NR	NR
Latex[1]	R	R	R	R	NR[2]	NR	NR[2]	NR
Chlorinated Rubber	R	R	R	R	R	NR	R	NR
Epoxy	R	R	R	R	R	R	R	NR
Epoxy Phenolic	R	R	R	R	R	R	R	R
Aggregate Filled Epoxy	R	R	R	R	R	R	R	R
Urethane Elastomers	R	R	R	R	R	R	R	R
Epoxy or Urethane Coal Tar	R	R	NR	R	R	R	R	R
Vinyl Ester/Polyester	R	R	NR	R	R	R	R	R

Reprinted with permission from NACE International. The complete edition of NACE Standard RP0591-91 is available from NACE International, P. O. Box 218340, Houston, Texas 77218-8340, phone: 713/492-0535, fax: 713/492-8254.
R = Recommended; NR = Not Recommended
[1]Excluding vinyl latices; [2]Certain latices may be suitable for service
NOTE: The recommendations provided are general. Candidate coating systems must be thoroughly evaluated to ensure that they are appropriate for the intended service conditions and meet other desired characteristics. The above list is not necessarily all-inclusive.
Courtesy of United States Army Corps of Engineers

STAIN REMOVAL

Stain removal can be accomplished with various methods. Brushing and washing are the usual ways to tackle a stain. Steam cleaning is frequently used to remove more stubborn stains. Additional cleaning options include water blasting, abrasive blasting, flame cleaning, mechanical cleaning, and chemical cleaning. Each method has its pros and cons.

Cleaning with water is usually not very invasive. It is done with a fine mist because too much pressure can drive stains deeper into concrete. The water is applied from the top down. When water alone is not getting the job done, brushing might be the answer. Soap, ammonia, and vinegar are sometimes used with water as a cleaning solution.

> All external tendons and tendon anchorage regions are required to be protected from corrosion.

Water blasting can remove some of the concrete surface. If too much pressure is used, it can be destructive. When dealing with dirt or chewing gum, steam cleaning is effective but expensive. Abrasive blasting is likely to remove some of the surface area of concrete. When this form of cleaning is used, the nozzle used to deliver the pressure should not be held too close to a concrete surface.

Organic stains may not be possible to remove with solvents. Flame cleaning is capable of cleaning these types of stains. However, the procedure may cause concrete to scale, and the fumes can be harmful.

Another way to remove stains is with mechanical equipment. This can range from a chisel to a grinder. Damage to concrete is not unusual when mechanical cleaning is performed. Many believe that chemical cleaning is the best bet for most types of stains. When used properly, the chemicals do not harm concrete. The downside to chemical cleaning is the health risks associated with the process. When chemical cleaning is used, close attention must be paid to proper procedures for working with the chemicals used.

> Unbonded construction that can be affected by repetitive loads must be protected from the possibility of fatigue in anchorages and couplers.

CLEANING DETAILS

The cleaning details for concrete depend on the type of stains being dealt with. Once you know the type of stain you are responsible for cleaning, the cleaning details for the job can be determined. Take iron rust as an example. Let's say that you have an iron rust stain that is light or shallow. In this situation, you should start by mopping the stained area with a solution of oxalic acid and water. Wait 2 to 3 hours, and scrub the surface with a stiff brush. Rinse as needed with clean water. This should do the trick, but if it doesn't, move to the next level of cleaning.

Deep stains can be treated with a poultice by mixing sodium citrate, glycerol, and diatomaceous earth or talc with water. Trowel the poultice over the stain. Leave the poultice in place for up to 3 days. If the stain is still evident, repeat the process. Just keep working it until the stain disappears.

Oil stains

Freshly spilled oil stains should not be rubbed. They should be soaked up with absorbent paper. Do not wipe the spill. Cover the spill area with an absorbent material. This could be something as simple as kitty litter, or it can be a more industrial absorbent. Wait a full day and then sweep up the absorbent material to remove the oil. Next, scrub the stained area with a scouring powder or strong soapy solution.

> Experimental and numerical analysis procedures are allowed when the procedures offer a reliable basis for design.

Old oil stains require a different method of attack. A mixture of equal parts of acetone and amyl acetate will be needed. This is mixed with some material, like flannel, to be laid over the stain. Once the cloth is over the stain, cover the cloth with a glass panel for about 15 minutes. Repeat the process as needed. When done, rinse the work area with clean water.

Grease

Grease can be scraped from a concrete surface. Scouring powder can be scrubbed on a grease stain, or you can scrub the area with a strong soap solution or sodium orthophosphate. If these methods fail, make a poultice of chlorinated solvents. Repeat the process as needed. Once the stain is under control, rinse the affected area with clean water.

> The specified compressive strength of concrete is 3,000 pounds per square inch.

Dirt

Dirt is normally pretty simple to remove from a concrete surface. Clean water is often all that is required. Soap and water may be needed in some cases. If you encounter difficult dirt stains, apply a solution of 19 parts water and 1 part hydrochloric acid to remove the stains. Steam cleaning can be used, but it can also be expensive. Dirt with a clay density can be removed with hot water that contains sodium orthophosphate and a scrub brush.

Mildew

Remove mildew stains with a mix of powdered detergent and sodium orthophosphate with commercial sodium hypochlorite solution and water. This is done by applying the mixture and waiting a few days. After the waiting period, use a brush to scrub the affected area. Rinse with clean water. Be careful when working around metal, since sodium hypochlorite may corrode metal.

The specified yield strength of nonprestressed reinforcement must not be in excess of 60,000 pounds per square inch.

Asphalt

Removing asphalt from a concrete surface can be a challenge. If the temperature is hot, start by chilling molten asphalt with ice. Scrap or chip off the asphalt while it is brittle, and then scrub the area with an abrasive powder and rinse the area completely with clean water. Applying solvents to emulsified asphalt can force the emulsions deeper into the concrete. Then scrub with scouring powder and rinse the area. You can use a poultice of diatomaceous earth or talc and a solvent to remove cutback asphalt. Once the poultice is dry, you should be able to brush it off.

Efflorescence

Water and a scrub brush will remove most new efflorescence stains. Older stains may require water blasting or sandblasting. Chemical removal with hydrochloric or phosphoric acid will generally work for difficult stains, but the solution can discolor the affected area. General practice calls for applying the chemical solution to all visible concrete in the cleaning zone to maintain a consistent appearance.

When the direction of reinforcement varies more than 10 degrees from the direction of principal tensile membrane force, an assessment must be done to prevent cracking.

Soot

Sometimes soot can be removed with water, scouring powder, and a scrub brush. A powdered pumice or grit can also be used for normal stains. Tough stains can be treated with trichloroethylene. This is done by soaking sections of cotton material in trichloroethylene and applying it to the stain. Horizontal stains can be covered with the saturated material and the material can be held in place with a heavy object. Vertical stains require the material to be braced in place against the stain.

Monitor the saturated material and remove it periodically. Wring the material out. Soak the material in fresh trichloroethylene and reapply the material to the stain. Continue the process until the stain is removed.

There is a health risk with handling trichloroethylene. Toxic gases may be created when the chemical is put into contact with fresh concrete or other strong alkalis. Use all proper safety precautions when working with any tools or chemicals.

COATINGS AND SEALING COMPOUNDS

Coatings and sealing compounds are applied to concrete surfaces to protect against chemical attacks. They are also used to control water penetration into concrete. Some thick coatings can be used to protect concrete from physical damage, but before any coating is used, it must be determined if the concrete needs protection. If a determination is made that a protective coating or sealant should be used, the concrete surface must be prepared properly. The surface must be sound, clean, and dry. The types of coatings and sealants available include the following:

- Silicones
- Siloxanes
- Silanes
- Cementitous coatings
- Urethanes
- Epoxy polyesters
- Latexes
- Chlorinated rubbers
- Epoxies
- Epoxy phenolics
- Aggregate-filled epoxies
- Thick film elastomers
- Thin-film polyurethane
- Latex
- Urethane elastomers
- Epoxy coal tar
- Urethane coal tar
- Vinylesters
- Polyesters

Once you have chosen the best coating or sealant for your needs, follow the manufacturer's recommendations for application. Surface temperatures and other variables come into play with coatings and sealants. This is why it is so important to read and follow the working instructions from each product manufacturer.

CHAPTER

Specialized repairs

8

There are many types of specialized concrete repairs. For example, a residential building contractor will generally think of foundation walls or concrete slabs. A civil engineer may think of a bridge deck or a lock wall. Architects might turn their thoughts to ornamental retaining walls. If you put your mind to it, you can come up with a wide variety of specialized concrete repairs.

Rehabilitation of existing concrete requires different tactics for different working conditions. A concrete pier submerged in water will not be repaired with the same procedures used for a parking deck. We are going to begin our discussion of specialized repairs with lock walls.

REHABBING LOCK WALLS

The rehabbing of lock walls can involve basic scaling to deeper damage. It is common to remove anywhere from 1 to 3 feet of concrete from the face of a lock wall. New Portland cement concrete is then used as a conventional concrete to resurface the wall. Other methods that can be considered are shotcrete, preplaced-aggregate concrete, and precise concrete that is in stay-in-place forms. Thin overlays are sometimes all that is required for a simple repair.

> When the dimensions of structural elements are evaluated, the assessment should be on critical sections.

CAST-IN-PLACE

Cast-in-place concrete can be economically feasible. Much of the cost factor is related to the depth of a repair. Shotcrete is a good option when repair sections range in thickness from 6 to 12 inches. Either of these two options is usually cost effective. Thicker sections of concrete require traditional concrete forming to maintain

doi: 10.1016/B978-1-85617-549-4.00008-3

the cost effectiveness. The minimum width that normally dictates the need for a form is 12 inches. Conventional cast-in-place concrete offers a number of advantages over other rehab materials:

- It can be proportioned to stimulate existing concrete substrate.
- It can minimize strains that can result from material incompatibility.
- Admixtures can be used in freezing and thawing temperatures.
- Conventional concrete uses proven methods that equipment and skilled workers are available for.

BLASTING LOCK WALLS

Blasting of lock walls is a common method of surface preparation before a repair is made. Workers often drill small-diameter holes along the top of a lock wall

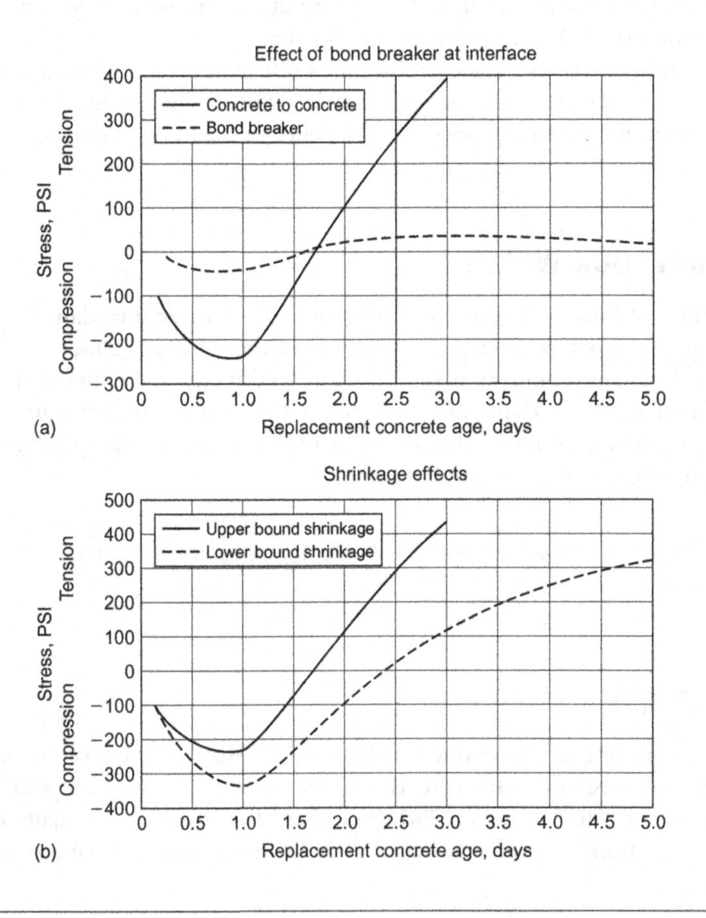

FIGURE 8.1

Results of a finite element analysis of a typical lock wall resurfacing.

Courtesy of United States Army Corps of Engineers

in a direction that is parallel to the removal face. Then a light explosive, such as a detonating cord, is loaded into the holes and cushioned by stemming the holes. Detonation is generally accomplished with electric blasting caps. Proper planning must go into this type of work. Engineers and other experts may use test results or historical reviews of previous jobs to determine a course of action for the blasting.

The code requires a set of final response measurements to be made within 24 hours of the time that a test load is selected for testing.

When blasting is complete, the remaining concrete must be checked for loose or defective concrete. This is done with sounding. When bad concrete is discovered, it can be removed with chipping, grinding, or water blasting. Before a repair is made, the concrete surface must be clean and free of debris that might compromise a bonding process.

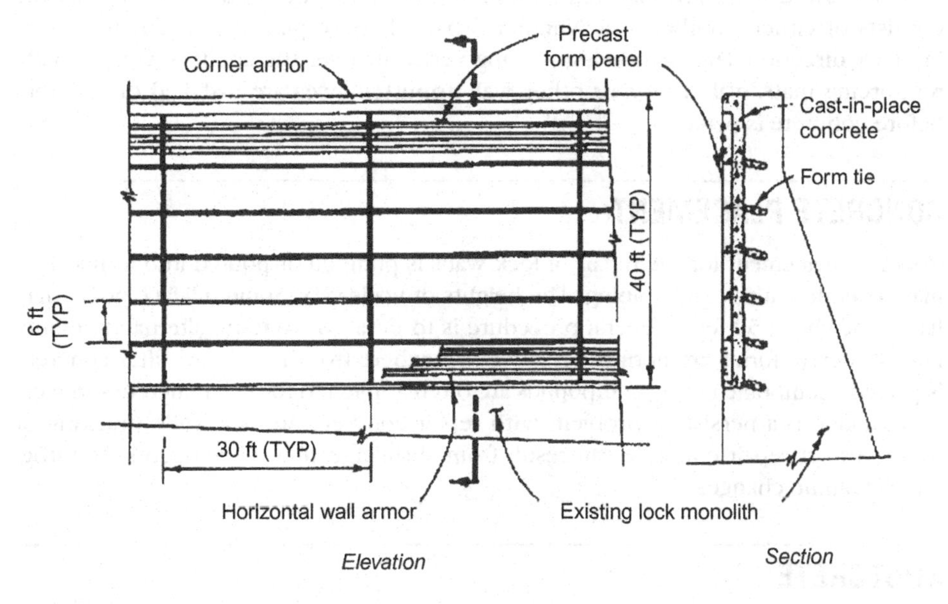

FIGURE 8.2

Precast concrete stay-in-place forming system for lock wall rehabilitation.

Courtesy of United States Army Corps of Engineers

ANCHORS

Anchors are needed for repairing the face of a lock wall. The type of anchors used can vary, but dowels are the typical choice. The dowels provide anchor points for reinforcing steel and for the bonding of new concrete to old concrete.

The spacing of dowels is determined by engineers or existing test data. A typical spacing for dowels is 4 feet apart. This is measured center to center. When openings will exist in a wall, the spacing of dowels should be no more than 2 feet apart.

Core samples are taken as a way of determining what types of repair materials will be best suited for the job. In the case of dowels, the core samples help to determine how deeply a dowel needs to be embedded in existing concrete. For example, a wall that has an average compressive strength of 3,000 psi, or more, requires dowels to be embedded to a depth that is at least 15 times the nominal diameter of the dowel. Field tests may prove that shorter holes can be used. It is also common practice to test a percentage of dowels installed to confirm a proper repair. One such requirement calls for 3 out of every 1,000 dowels be tested.

> The code contains special provisions for seismic designs.

Once the dowels are installed, reinforcing steel can be installed. This normally consists of either Number 5 or Number 6 bars that are placed on 12-inch centers in every direction. The reinforcing is hung vertically over the anchors. Occasionally, reinforcing mats, wall armor, or other wall appurtenances are installed on anchors before concrete is placed.

CONCRETE PLACEMENT

Concrete placement for the repair of lock walls is pumped or poured into forms. This may be done with flexible piping. The heights of walls vary. Some full-face pours may have a height of 50 feet. General procedure is to pour concrete on alternating monoliths. Concrete forms are normally removed anywhere from 1 to 3 days after concrete is placed. Membrane curing compounds are often applied to formed concrete surfaces.

Cracking is a persistent problem with repair concrete on lock walls. Cracking of thin layers of repair concrete can result from shrinkage, thermal gradients, or autogenous volume changes.

SHOTCRETE

Shotcrete is a cost-effective way of repairing walls when the repair thickness does not exceed 6 inches. Overall, shotcrete is a durable material that does an adequate job of structural repair. There are, however, some potential problems associated with the material:

- Moisture may be trapped between the existing lock wall and the shotcrete repair. This can be a serious problem when freezing and thawing conditions exist.
- Spalling can happen over time.
- Delamination may occur.

PREPLACED-AGGREGATE CONCRETE

Preplaced-aggregate concrete is more resistive to shrinkage and creep than conventional concrete is. This is due to the aggregate. The net result is more protection against cracking. This type of concrete can be used on numerous types of structures. The cost involved with using preplaced-aggregate concrete is likely to be more, but it may be worth it in the long run.

> When longitudinal reinforcement is required by design, the welding of stirrups, ties, inserts, and similar elements is not allowed.

FIGURE 8.3

Precast concrete construction.

Copyright © Gary S. Figallo. Courtesy of Faddis Concrete Products

PRECAST CONCRETE

Precast concrete has a lot to offer in concrete repairs. When compared to cast-in-place concrete, the advantages of using precast concrete are numerous:

- Minimal cracking
- Durability
- Rapid construction
- Lower maintenance costs
- Improved appearance
- Lower impact from onsite weather conditions
- The ability to inspect the concrete before it is placed in use
- May eliminate the need for dewatering a lock chamber during repairs

CUTOFF WALLS

Concrete cutoff walls are cast-in-place structures. They are used to provide a positive cutoff of the flow of water under or around a hydraulic structure, such as a dam. Geotechnical monitoring and review prior to making a decision to build a cutoff wall is prudent.

The average cutoff wall is between 2 to 4 feet thick. The procedure for preparing for and creating a cutoff wall can be tricky. A concrete-lined guide trench is made along the axis of the wall to be repaired. This type of trench is usually only a few feet deep. This gives a working service on both sides of the wall and helps to maintain the alignment of the wall.

> Plain concrete is not allowed for use in footings on piles.

Concrete is placed in the trench to create the cutoff wall. Discontinuities in the concrete can cause serious performance problems and must be screened for. The concrete mixture is very important. It will be used for tremie placement, and this requires strict adherence to required specifications. You are not looking for compressive strength. The key elements are flowability and cohesion.

Test panels should be created in noncritical locations near the wall location. These panels can then be tested to determine if the concrete mixture and the placement process is providing the desired result.

PRECAST CONCRETE APPLICATIONS

Precast concrete has been used extensively in recent years. We have already discussed the advantages to this type of repair process. Here are some of the types of concrete structures where precast-concrete can be used:

- Navigation locks
- Dams
- Channels
- Floodwalls
- Levees
- Coastal structures
- Marine structures
- Bridges
- Culverts
- Tunnels
- Retaining walls
- Noise barriers
- Highway pavement

FIGURE 8.4

Precast concrete retaining wall.

Copyright © Gary S. Figallo. Courtesy of Faddis Concrete Products

UNDERWATER REPAIRS

Underwater repairs are sometimes a necessity. The cost of dewatering can be extremely expensive. As you might imagine, an underwater repair requires special tactics and techniques. Naturally, the surface to be repaired must be clean. Underwater, specialized equipment is essential for these types of repairs.

Common methods to excavate underwater include air lifting, dredging, and jetting. Air lifts are often used when the repair is in water up to 75 feet deep. When greater depths are encountered, dredging and jetting are the prime options.

> A strut that is wider at midlength than at its ends is known as a bottle-shaped strut.

Debris removal is normally required. This can involve the removal of cobbles, sediment, or reinforcing steel. Cobbles and sediment can be dislodged and removed with basic excavation techniques. Steel is another matter. It is usually removed with one of two methods: mechanical or thermal. Mechanical removal is often accomplished with either hydraulically powered shears or bandsaws.

There are three potential thermal techniques for underwater cutting: oxygen-arc cutting, shielded-metal-arc cutting, and gas cutting. Most people prefer oxygen cutting. A newer type of cutting that is available and evolving is abrasive-jet cutting.

Cleaning is always of prime importance when working with concrete repairs. This is no different from when working underwater. The hard way uses hand tools, such as chisels, brushes, abrasive disks, and so forth. Powered cleaning tools make the job easier, and on large jobs, a self-propelled cleaning vehicle can be used.

> An abrupt change in geometry or loading is discontinuity.

When mixing concrete for submerged use, the mixture must be workable, cohesive, and protected from water until it is in place. A tremie pipe has long been used for underwater concrete placement. Vertical placement has generally been done with the tremie method. However, tremie pipe has been placed at up to a 45-degree angle to slow the flow of concrete. This is due to the fact that some straight drops of free-falling concrete are not always desirable.

Another placement method is known as the Hydrovalve method and the Kajima Double Tube tremie method. These are both variations of the traditional tremie method. A flexible hose that collapses under hydrostatic pressure and carries a controlled amount of concrete down the hose in slugs is used. This helps to prevent segregation. The methods are reliable, inexpensive, and can be used by any contractor who has workers who are skilled at working underwater.

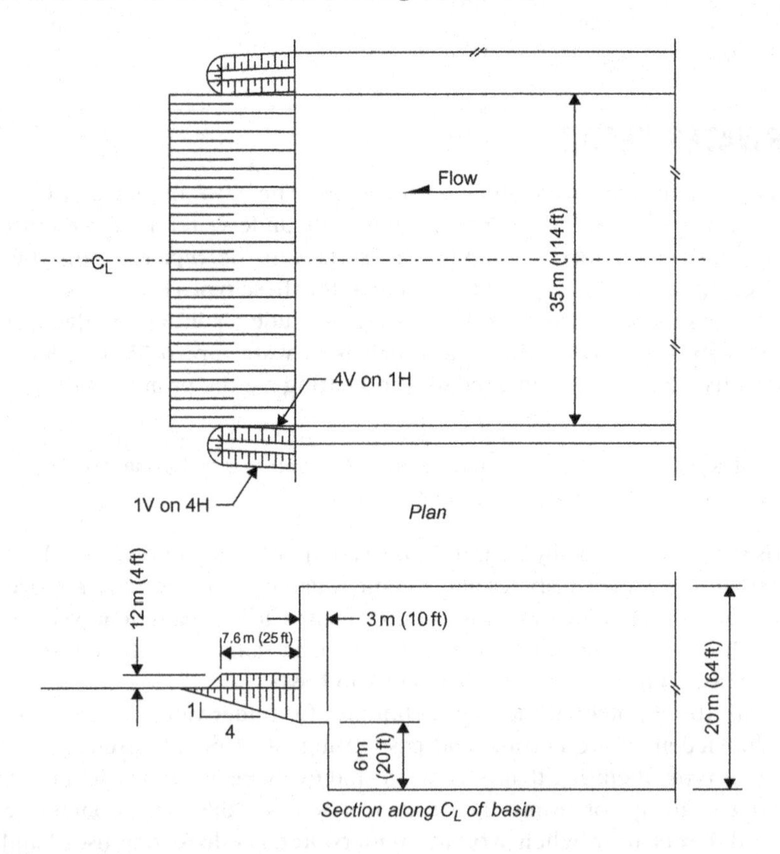

FIGURE 8.5

Stilling basin wall extension fills at Dworshak dam.

Courtesy of United States Army Corps of Engineers

Pumping is frequently preferred over the tremie method for placing concrete below water. When thin layers are needed, pumping offers multiple advantages: (1) there are fewer transfer points, (2) gravity-feeding problems are eliminated, and (3) being able to use a boom for placement allows for better control.

If you encounter a large void that needs to be filled underwater, consider using preplaced-aggregate concrete. The concrete should be placed in a form, and then grout should be injected from the bottom of the preplaced aggregate.

You must prevent loss of material from the top of the concrete forms. This is normally done by placing a permeable fabric next to the concrete, backed with a wire mesh. This is supported by a stronger backing of perforated steel and plywood. Pressure is put out as grout is injected. This can raise the concrete forms. Dowels can be used to protect against excessive form movement.

Precast concrete panels and modular sections have been used in underwater repairs. They have been used to repair dams, stilling basins, and lock walls. Successful underwater repairs and concrete forms have been made with prefabricated steel panels. These same types of panels have been used to control erosion. Both of these options have their pros and cons. Consider the underwater application at hand, and weigh both options when choosing an appropriate repair procedure.

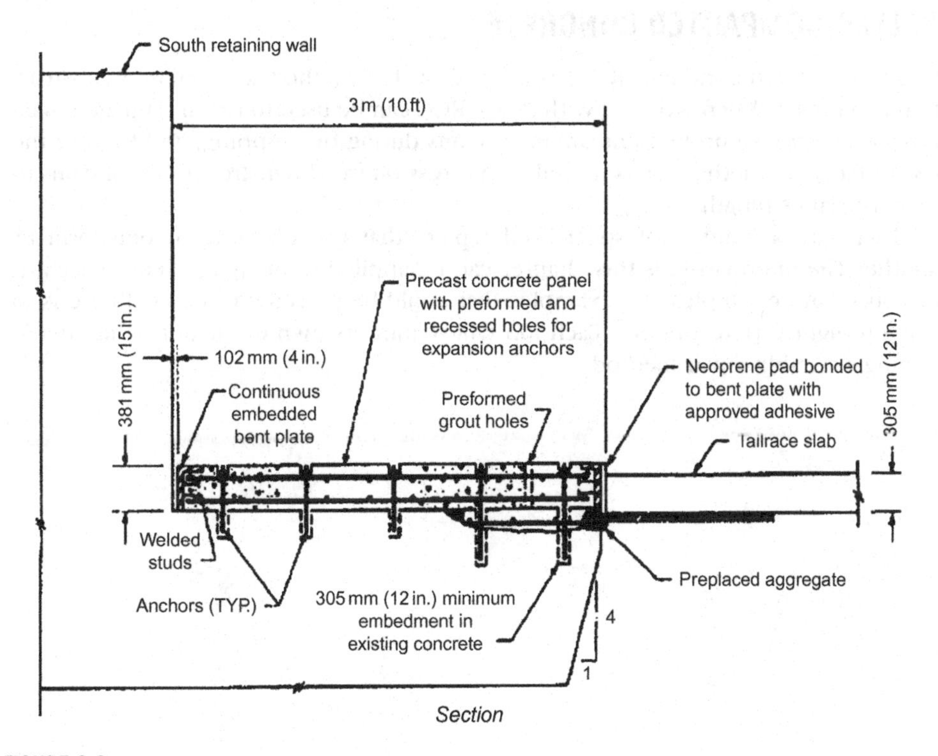

FIGURE 8.6

Underwater repair of concrete spalling at Gavins Point dam.

Courtesy of United States Army Corps of Engineers

GEOMEMBRANE WORK

Geomembrane work is very useful when dealing with dams. Seepage control is just one use for geomembranes. They can be used when working with canals, reservoirs, storage basins, dams, and tunnels. In Europe, geomembranes have been used to repair old concrete and masonry dams.

A geomembrane is a polymeric membrane that constitutes a flexible, watertight material with a thickness of one-half millimeter or more. The manufacture of geomembranes is done with a wide range of polymers. The types of polymers used can include plastics, elastomers, and blends of polymers.

In the old days, geomembranes were installed on the upstream side of structures with either nails or adhesives. More recently, stainless steel anchors are prolific. This type of system uses two vertical U-shaped anchors. One anchor is larger than the other. The smaller anchor is fastened to the face of the structure first. Then the larger anchor is placed over the smaller one and connected to the face. This creates two voids to be filled with geomembranes.

ROLLER-COMPACTED CONCRETE

Roller-compacted concrete (RCC) is a low-cost, fast method for concrete construction and repair. When working with dams, RCC can be used to repair damaged overflow structures, to protect embankment dams during overtopping, and to raise the crest of a dam. Another use is to build a buttress on the downstream side of dams to increase dam strength.

There are a number of specialized repairs that use concrete in one form or another. The information is this chapter can be applied to many of them. There are, of course, more samples and examples that could be presented. The goal here is to bring to light various options. Each job will require its own evaluation when determining a suitable repair method.

CHAPTER

Troubleshooting defects in concrete

9

Troubleshooting defects in concrete is required to make a proper repair. We have already discussed examples of this. Here we are going to concentrate on the troubleshooting process. Some defects can be difficult to troubleshoot. The elements of the defect might mimic those of one cause while they are, in fact, due to a different cause. This can lead to an improper repair. Knowing what to look for and where to look for it can save you a lot of time, money, and frustration.

TOO MUCH WATER

What do you think is the most common cause of concrete failure? It may surprise you to learn that it is too much water in the concrete mix. When excessive water is added to the mix, many conditions can occur, including the following:

- Reduced strength
- Reduced abrasion resistance
- Increased curing time
- Increased drying shrinkage
- Increased porosity
- Increased creep

When concrete is installed with excessive water in the mixture, it may have to be removed and replaced. A repair may not be feasible. If the damage is shallow—say, less than 1.5 inches—a concrete sealer could offer a repair option. If a sealant is used, expect to reapply the sealer on a routine basis. Repairs to concrete damaged by excessive water in the mix are sometimes able to extend the useful life of the concrete, but they are rarely a permanent solution.

Damage that is 1.5 to 6 inches deep might be able to be repaired with an epoxy-bonded replacement concrete. Deeper flaws can be filled with replacement concrete.

doi: 10.1016/B978-1-85617-549-4.00009-5

The best bet is to monitor concrete mixes as they are being used to avoid these problems.

BAD DESIGN DATA

Bad design data can result in a host of concrete problems. This problem can affect any element of an installation or repair. We would have to cover every potential defect to list all of the risks of bad designs. However, there are some design and or installation factors to consider as common causes of damaged concrete.

How many times have you seen cracks, humps, flaking, or similar surface defects located near encased conduits and pipes? If electrical conduits or outlets are placed in a way that will have them close to the concrete surface, you can expect problems. The same is true when the bases of handrails and similar surface-mounted elements are placed too near the exterior corners of concrete surfaces.

Some pipe used for handrails can experience thermal expansion and contraction. You may wonder what this has to do with concrete. Since the handrails are secured to the concrete, stress from the movement of the handrail can damage the concrete. Slip joints need to be installed in the concrete to absorb this movement without cracking. Once cracking occurs, moisture can enter the surface, and where freezing and thawing is a factor, the cracks and damage will accelerate.

A major cause of concrete damage on a bridge or hydroelectric structure is insufficient cover over reinforcing steel. It is normally acceptable to add a 3-inch layer of new concrete over an affected area. However, if corrosion is probable, the thickness of the repair layer should be not less than 4 inches.

Concrete slabs need to be equipped with expansion and contraction joints. If they are not installed, cracking is likely. This can be a very big problem for bridge decks, dam roadways, floors of spillways, and such. Delamination is common under these conditions. Bad designs are bound to lead to problems. Repairs are costly. Try to ensure good design to avoid a multitude of potential problems.

CHEMICAL ATTACK

Concrete can be damaged from chemical attacks. Sodium, magnesium, calcium sulfates, and salts that can be found in alkali soils and groundwater could cause expensive concrete damage. A chemical reaction can occur between these natural elements and the hydrated line and hydrated aluminate in cement paste. Calcium sulfate and calcium sulfoaluminate may form. When all of this occurs, concrete may experience expansion damage. Type V Portland cement has a low calcium aluminate content and is very resistant to sulfate reaction.

Concrete that is being damaged by chemical reactions can benefit from a thin polymer concrete overlay. Wetting and drying the concrete can slow the rate of deterioration. When damage gets out of control, the concrete should be replaced with type V Portland cement.

ALKALI-AGGREGATE REACTION

Alkali-aggregate reaction can damage concrete. The cause of this type of reaction can be related to sand and aggregates, such as opal, chert, flint, and volcanic that have a high concentration of high silica. Calcium, sodium, and potassium hydroxide alkalies in Portland cement could have a reaction with these components. If a reaction occurs, it can result is destructive expansion. A low-alkali Portland cement and fly ash pozzolan used in new construction can reduce this risk.

Once a reaction has begun damaging concrete, repairs are futile until all of the damage is done. In some cases, cutting relief slots in the concrete can extend the useful life of a structure. In many cases the destruction will cease after enough time has passed. When the damage is no longer active, normal repairs can be made.

FREEZING

Repeated freezing and thawing damages a lot of concrete. Concrete sealers can be used to minimize the risk of this type of destruction. The key to preventing damage from freezing and thawing is protecting concrete from excessive moisture content. When a repair is needed, it is normally done with replacement concrete.

MOVING WATER

Moving water that is channeled by concrete structures can contain destructive elements. Abrasion from waterborne sand and rocks can do a number on a concrete surface. Generally, the water must be moving quickly for this to be a significant problem. You will know this type of defect by the polished appearance of the concrete surface. When repairs are required, they are usually made with either polymer concrete or silica fume concrete.

CAVITATION

Cavitation can damage concrete when fast-moving water encounters discontinuities on the flow surface. When flowing water is lifted off the concrete surface, bubbling can occur. This creates a negative pressure zone. When the bubbles come into contact with concrete and burst, there is a very high-pressure impact. It may seem strange that bubbles can damage concrete, but they can. As the bubbles pop, they can remove particles from concrete. Over time, this damage can grow to serious levels. Common locations for this type of damage are water control gates and gate frames. This type of damage is so intense that not even cast iron or stainless steel can stand up against it. Repair of cavitation damage normally calls for total replacement of the damaged concrete. There are, however, some situations where a repair can be made with an epoxy-bonded replacement.

THE ROUNDUP

The roundup of all possible concrete defects is not feasible in the space available here. We have covered the types of damage, causes, and repairs for which you will normally require troubleshooting skills. If concrete is suffering from damage of being overloaded, the damage is usually obvious and noted over time. Cracking can come from a variety of causes. Corrosion in reinforcing steel is not considered a cause for concrete damage but rather the result of damage from some relevant cause.

When you are troubleshooting a cause of damage for repair options, be thorough. Consider all viable options before making a repair decision. Attempting to correct a problem with the wrong repair procedure is simply going to cause more trouble and cost more money. Time spent in evaluation and research is time well spent.

CHAPTER

Worksite safety

10

Worksite safety doesn't get as much attention as it should. Far too many people are injured on jobs every year. Most of the injuries could be prevented, but they are not. One of the main reasons for this is that people are in a hurry to make a few extra bucks, so they cut corners. This happens with employees, contractors, and piece workers. It even affects hourly installers who want to shave 15 minutes off their workday so they can head back to the shop early.

Based on my field experience, most accidents occur as a result of negligence. Workers try to take shortcuts, and they wind up getting hurt. This has proved true with my personal injuries. I've only suffered two serious on-the-job injuries, and both of them were a direct result of my carelessness. I knew better than to do what I was doing when I was hurt, but I did it anyway. Well, sometimes you don't get a second chance, and the life you affect may not be your own. So let's look at some sensible safety procedures that you can implement in your daily activity.

A DANGEROUS JOB

Construction can be a very dangerous trade. The tools of the trade have the potential to be killers. Requirements of the job can place you in positions where a lack of concentration could result in serious injury or death. The fact that working with concrete can be dangerous is no reason to rule out the trade as your profession. Driving can be extremely dangerous, but few people never get behind the wheel out of fear.

Fear is generally a result of ignorance. When you have a depth of knowledge and skill, fear begins to subside. As you become more accomplished at what you do, fear is forgotten. While it is advisable to learn to work without fear, you should never work without respect. There is a huge difference between fear and respect.

If, as an installer, you are afraid to climb up high enough to set a pour form, you are not going to last long in the plumbing trade. However, if you scurry up recklessly,

doi: 10.1016/B978-1-85617-549-4.00010-1

you could be injured severely, perhaps even killed. You must respect the position you are putting yourself in. If you are using a ladder, you must respect the outcome of what a mistake could have.

Being afraid of heights could limit or eliminate your career. Respect is the key. If you respect the consequences of your actions, you are aware of what you are doing, and your odds for a safe result improve.

> Safe working conditions are a good place to start when building a career in the concrete business. If you are injured, your ability to do your job is likely to be limited.

Many young installers are fearless in the beginning. They think nothing of darting around on a roof or jumping down in a trench. As their careers progress, they usually hear about or see on-the-job accidents. Someone gets buried in a cave-in of a trench. Somebody falls off a roof. A metal ladder being set up hits a power line. The list of possible job-related injuries is a long one.

Millions of people are hurt every year in job-related accidents. Most of these people were not following solid safety procedures. Sure, some of them were victims of unavoidable accidents, but most were hurt by their own hand, in one way or another. You don't have to be one of these statistics.

In over 30 years in construction work, I have only been hurt seriously on the job twice. Both times, I had only myself to blame. I got careless. One time, I let loose clothing and a powerful drill work together to chew up my arm. The other time, I tried to save myself the trouble of repositioning my stepladder while drilling holes in floor joists. My desire to save a minute cost me torn stomach muscles and months of pain from a twisting drill.

My accidents were not mistakes; they were *stupidity*. Mistakes are made through ignorance. I wasn't ignorant of what could happen to me. I knew the risk I was taking, and I knew the proper way to perform my job. Even with my knowledge, I slipped up and got hurt. Luckily, both of my injuries healed, and I didn't pay a lifelong price for my stupidity.

During my long career, I have seen a lot of people get hurt. Most of these people have been helpers and apprentices. Of all the on-the-job accidents I have witnessed, every one of them could have been avoided. Many of the incidents were not extremely serious, but a few were. As a concrete worker, you will be doing some dangerous work. Hopefully, your employer will provide you with quality tools and equipment. If you have the right tool for the job, you are off to a good start in staying safe.

> Safety training is something you should seek from your employer. Some contractors fail to tell their employees how to do their jobs safely. It is easy for someone who knows a job inside and out to forget to inform an inexperienced person of potential danger.

For example, a supervisor might tell you to break up the concrete around a pipe to allow the installation of new plumbing and never consider telling you to wear safety glasses. The supervisor will assume you know that the concrete is going to fly up in your face as it is chiseled up. However, as a rookie, you might not know about the reaction concrete has when hit with a cold chisel. One swing of the hammer could cause extreme damage to your eyesight.

Simple jobs, like the one in the example, are all it takes to ruin a career. You might be really on your toes when asked to scoot across an I-beam, but how much thought are you going to give to carrying a few bags of concrete mix to a mixer? The risk of falling off the I-beam is obvious. Hurting your back by carrying heavy loads the wrong way may not be so obvious. Either way, you can have a work-stopping injury.

Safety is a serious issue. Some job sites are very strict in the safety requirements maintained. But a lot of jobs have no written rules of safety. If you are working on a commercial job, supervisors are likely to make sure you abide by the rules of the Occupational Safety and Health Administration (OSHA). Failure to comply with OSHA regulations can result in stiff financial penalties. However, if you are working residential jobs, you may never work on a job where OSHA regulations are observed.

In all cases, you are responsible for your own safety. Your employer and OSHA can help you to remain safe, but in the end, it is up to you. You are the one who has to know what to do and how to do it. And not only do you have to take responsibility for your own actions, you also have to watch out for the actions of others. It is not unlikely that you could be injured by someone else's carelessness. Now that you have had the primer course, let's get down to the specifics of job-related safety.

As we move into specifics, you will find the suggestions in this chapter broken down into various categories. Each category will deal with specific safety issues related to the category. For example, in the section on tool safety, you will learn procedures for working safely with tools. As you move from section to section, you may notice some overlapping of safety tips. For example, in the section on general safety, you will see that it is wise to work without wearing jewelry. Then you will find jewelry mentioned again in the tool section. The duplication is done to pinpoint definite safety risks and procedures. We will start into the various sections with general safety.

GENERAL SAFETY

General safety covers a lot of territory. It starts from the time you get into the company vehicle and carries you right through to the end of the day. Much of the general safety recommendations involve the use of common sense. Now, let's get started.

> If you are unloading heavy items, don't put your body in awkward positions. Learn the proper ways for lifting, and never lift objects inappropriately.

Vehicles

Many construction workers are given company trucks for their use in getting to and from jobs. You will probably spend a lot of time loading and unloading company trucks. And, of course, you will spend time either riding in or driving them. All of these areas can threaten your safety.

If you will be driving the truck, take the time to get used to how it handles. Loaded work trucks don't drive like the family car. Remember to check the vehicle's fluids, tires, lights, and related equipment. Many company trucks are old and have seen better days. Failure to check the vehicle's equipment could result in unwanted headaches. Also remember to use the seat belts; they *do* save lives.

Apprentices are normally charged with the duty of unloading the truck at the job site. There are a lot of ways to get hurt in doing this job. Many trucks use roof racks to haul supplies and ladders. If you are unloading these items, make sure they will not come into contact with low-hanging electrical wires. Aluminum ladders make very good electrical conductors, and they will carry the power surge through you on the way to the ground. If you are unloading heavy items, don't put your body in awkward positions. Learn the proper ways for lifting, and never lift objects inappropriately. If the weather is wet, be careful climbing on the truck. Step bumpers get slippery, and a fall can impale you on an object or bang up your knee.

When it is time to load the truck, observe the same safety precautions you did in unloading. In addition to these considerations, always make sure your load is packed evenly and well secured. Be especially careful of any load you attach to the roof rack, and always double-check the cargo doors on trucks with utility bodies. If you are carrying a load of forms in the bed of your truck, make very sure that they are strapped in securely.

Eye and ear protection

Eye and ear protection is often overlooked. An inexpensive pair of safety glasses can prevent you from permanently losing your sight. Ear protection reduces the effect of loud noises, such as jackhammers and drills. You may not notice much benefit now, but in later years you will be glad you wore it. If you don't want to lose your hearing, wear ear protection when subjected to loud noises.

Clothing

Clothing is responsible for a lot of on-the-job injuries. Sometimes it is the lack of clothing that causes the accidents, and there are many times when too much clothing creates the problem. Generally, it is wise not to wear loose-fitting clothes. Shirttails should be tucked in, and short-sleeve shirts are safer than long-sleeved shirts when operating some types of equipment.

Caps can save you from minor inconveniences, and hard hats provide some protection from potentially damaging accidents, like having a steel fitting dropped on your head. If you have long hair, keep it up and under a hat.

Good footwear is essential in the trade. Normally a strong pair of hunting-style boots will be best. The thick soles provide some protection from nails and other sharp objects you may step on. Boots with steel toes can make a big difference in your physical well being. If you are going to be climbing, wear foot gear with a flexible sole that grips well. Gloves can keep your hands warm and clean, but they can also contribute to serious accidents. Wear gloves sparingly, depending on the job you are doing.

Jewelry

On the whole, jewelry should not be worn in the workplace. Rings can inflict deep cuts in your fingers. They can also work with machinery to amputate fingers. Chains and bracelets are equally dangerous, probably more so.

Kneepads

Kneepads can make your job more comfortable, and protect your knees. Some workers spend a lot of time on their knees, and pads should be worn to ensure that they can continue to work for many years.

The embarrassment factor plays a significant role in job-related injuries. People, especially young people, feel the need to fit in and to make a name for themselves. It is no secret that construction workers often consider themselves super human beings. The work can be hard, and doing it has the side benefit of making you stronger. But you can't allow safety to be pushed aside for the purpose of making yourself look invulnerable.

Too many people believe that working without safety glasses, ear protection, and so forth makes them tougher. That's just not true; it may make them dumber, and it may land them in the hospital, but it does not make them look stronger.

Don't fall into the trap so many young tradespeople do. Never let people goad you into bad safety practices. Some people are going to laugh at your kneepads. Let them laugh. You will be the one with great knees when they are hobbling around on canes. I'm dead serious about this issue. There is nothing sissy about safety. Wear your gear in confidence, and don't let the few jokesters get to you.

TOOL SAFETY

Tool safety is a big issue. Anyone in the trades will work with numerous tools. All of these tools are potentially dangerous, but some of them are especially hazardous.

This section is broken down by the various tools used on the job. You cannot afford to start working without the basics in tool safety. The more you can absorb on tool safety, the better off you will be.

The best starting point is reading all the literature available from the manufacturers of your tools. The people who make the tools provide some good safety suggestions with them. Read and follow the manufacturers' recommendations.

The next step in working safely with your tools is to ask questions. If you don't understand how a tool operates, ask someone to explain it to you. Don't experiment on your own, or the price you pay could be much too high.

Common sense is irreplaceable in the safe operation of tools. If you see an electrical cord with cut insulation, you should have enough common sense to avoid using it. In addition to this type of simple observation, you will learn some interesting facts about tool safety. Now, let me tell you what I've learned about tool safety over the years.

> Know your tools well and keep them in good repair. Be especially sure to check all electrical cords for damage.

There are some basic principles to apply to all of your work with tools. We will start with the basics, and then move on to specific tools:

- Keep body parts away from moving tool parts.
- Don't work with poor lighting conditions.
- Be careful of wet areas when working with electrical tools.
- If special clothing is recommended for working with your tools, wear it.
- Use tools only for their intended purposes.
- Get to know your tools well.
- Keep your tools in good condition.

Now, let's take a close look at the tools you are likely to use.

Drills and bits

Drills have been my worst enemy. The two serious injuries I have received were both related to my work with a drill. The drills most construction workers use are not the little pistol-grip, handheld types of drills most people think of. The day-to-day drilling done in concrete work involves the use of large, powerful drills. These drills have enormous power when they get in a bind. Hitting an obstruction while drilling can do a lot of damage. You can break fingers, lose teeth, suffer head injuries, and a lot more. As with all electrical tools, you should always check the electrical cord before using your drill. If the cord is not in good shape, don't use the drill.

Always know what you are drilling into. If you are doing new-construction work it is fairly easy to look before you drill. However, drilling in a remodeling job can be much more difficult. You cannot always see what you are getting into. If you are unfortunate enough to drill into a hot wire, you can get a considerable electrical shock.

The bits you use in a drill are part of the safe operation of the tool. If your drill bits are dull, sharpen them. Dull bits are much more dangerous than sharp ones. If you will be drilling metal, be aware that the metal shavings will be sharp and hot.

Power saws

Concrete installers don't use power saws as much as carpenters, but they do use them. The most common types of power saws used by concrete workers are concrete saws, reciprocating saws, and circular saws. These saws are used to cut concrete, form material, pipe, plywood, floor joists, and a whole lot more. All of the saws have the potential for serious injury.

Reciprocating saws are reasonably safe. Most models are insulated to help avoid electrical shocks if a hot wire is cut. The blade is typically a safe distance from the user, and the saws are pretty easy to hold and control. However, the brittle blades do break, and this could result in an eye injury.

Circular saws are used by concrete workers occasionally. The blades on these saws can bind and cause the saws to kick back. If you keep your body parts out of the way and wear eye protection, you can use these saws safely. Concrete saws are heavy and noisy, and they generate a lot of dust. Protect your eyes and your respiratory system from the flying chips of concrete and dust.

Power mixers

Power mixers are often used on small jobs. These tools make the mixing of concrete simple and require far less effort than what would be needed to mix the concrete manually. Whenever there are moving parts, as there are in mixers, there are safety risks. Workers must be careful not to get body parts or clothing caught in the moving parts of tools. Also, keep your back in mind when lifting bags of material to dump into a mixer.

Air-powered tools

Air-powered tools are not used often by concrete workers. Jackhammers are probably the most used air-powered tools for individuals rehabbing concrete. When using tools with air hoses, check all connections carefully. If you experience a blowout, the hose can spiral wildly out of control. The air hose can also create a tripping hazard that must be avoided. Any type of power washer, sandblaster, or related equipment can cause injuries.

Powder-actuated tools

Powder-actuated tools are used to secure objects to hard surfaces, like concrete. If the user is properly trained, these tools are not too dangerous. However, good training, eye protection, and ear protection are all necessary. Misfires and chipping hard surfaces are the most common problems with these tools.

Ladders

Both stepladders and extension ladders are used frequently by construction workers. Many ladder accidents are possible. You must always be aware of what is around you when handling a ladder. If you brush against a live electrical wire with a ladder you are carrying, your life could be over. Ladders often fall over when the people using them are not careful. Reaching too far from a ladder can be all it takes to fall.

Rebar that is placed prior to a concrete pour should be installed with protector cups to reduce the risk of injury if someone falls onto the reinforcement bars.

When you set up a ladder or a rolling scaffold, make sure it is set up properly. The ladder should be on firm footing, and all safety braces and clamps should be in place. When using an extension ladder, many plumbers use a rope to tie rungs together where the sections overlap. The rope provides an extra guard against the ladder's safety clamps failing and the ladder collapsing. When using an extension ladder, be sure to secure both the base and the top.

I had an unusual accident on a ladder. I was on a tall extension ladder, working near the top of a commercial building. The top of my ladder was resting on the edge of the flat roof. There was metal flashing surrounding the edge of the roof, and the top of the ladder was leaning against the flashing. There was a picket fence behind me and electrical wires entering the building to my right. The entrance wires were a good ways away, so I was in no immediate danger. As I worked on the ladder, a huge gust of wind blew around the building. I don't know where it came from; it hadn't been very windy when I went up the ladder. The wind hit me and pushed both me and the ladder sideways. The top of the ladder slid easily along the metal flashing, and I couldn't grab anything to stop myself. I knew the ladder was going to go down, and I didn't have much time to make a decision. If I pushed off of the ladder, I would probably be impaled on the fence. If I rode the ladder down, it might hit the electrical wires and fry me. I waited until the last minute and jumped off of the ladder.

I landed on the wet ground with a thud, but I missed the fence. The ladder hit the wires and sparks flew. Fortunately, I wasn't hurt, and electricians were available to take care of the electrical problem. This was a case where I wasn't really negligent, but I could have been killed. If I had secured the top of the ladder, none of that would have happened.

Screwdrivers and chisels

Eye injuries and puncture wounds are common when working with screwdrivers and chisels. When the tools are used properly and safety glasses are worn, few accidents occur. The key to avoiding injury with most hand tools is simply to use the right tool for the job. If you use a wrench as a hammer or a screwdriver as a chisel, you are asking for trouble.

There are, of course, other types of tools and safety hazards found in the concrete trade. However, this list covers the ones that result in many injuries. In all cases, observe proper safety procedures and utilize safety gear, such as eye and ear protection.

COWORKER SAFETY

Coworker safety is the last segment of this chapter. I am including it because workers are frequently injured by the actions of coworkers. This section is meant to protect you from others and to make you aware of how your actions might affect your coworkers.

Most installers find themselves working around other people. This is especially true on construction jobs. When working around other people, you must be aware of their actions as well as your own. If you are walking out of a house to get something off the truck and a roll of roofing paper gets away from a roofer, you could get an instant headache.

If you don't pay attention to what is going on around you, it is possible to wind up in all sorts of trouble. Cranes lose their loads sometimes, and such a load landing on you is likely to be fatal. Equipment operators don't always see the concrete worker kneeling down to drive in a section of rebar. It's not hard to have a close encounter with heavy equipment. While we are on the subject of equipment, let me bore you with another war story.

One day I was in a ditch. The section of ditch that I was working in was only about four feet deep. There was a large pile of dirt near the edge of the trench; it had been created when the ditch was dug. The dirt wasn't laid back like it should have been; it was piled up. As I worked in the ditch, a backhoe came by. The operator had no idea I was in the ditch. When he swung the backhoe around to make a turn, the small scorpion-type bucket on the back of the equipment hit the dirt pile.

I had stood up when I heard the hoe approaching, and it was a good thing I had. When the equipment hit the pile of dirt, part of the mound caved in on me. I tried to run, but it caught both of my legs and the weight drove me to the ground. I was buried from just below my waist down. My head was okay, and my arms were free. I was still holding my shovel.

I yelled, but nobody heard me. I must admit, I was a little panicked. I tried to get up but couldn't. After a while, I was able to move enough dirt with the shovel to crawl out from under the dirt. I was lucky. If I had been on my knees working, I would have been smothered. As it was, I came out of the ditch no worse for wear. But, boy, was I mad at the careless backhoe operator. (I won't go into the details of the little confrontation I had with him.)

That accident is a prime example of how other workers can hurt you and never know they did it. You have to watch out for yourself at all times. As you gain field experience, you will develop a second nature for impending coworker problems. You will learn to sense when something is wrong or is about to go wrong. But you have to stay alive and healthy long enough to get that experience.

Always be aware of what is going on over your head. Avoid working under other people and hazardous overhead conditions. Let people know where you are, so you won't get stranded on a roof or in an attic when your ladder is moved or falls over.

You must also remember that your actions could harm your coworkers. If you are on a roof to flash a pipe and your hammer gets away from you, somebody could get hurt. Open communication between workers is one of the best ways to avoid injuries. If everyone knows where everyone else is working, injuries are less likely. Primarily, ***think***, and then think some more. There is no substitute for common sense. Try to avoid working alone, and remain alert at all times.

CHAPTER

First aid

11

Everyone should invest some time in learning the basics of first aid. You never know when having skills in first-aid treatments may save your life. Plumbers live what can be a dangerous life. On-the-job injuries are not uncommon. Most injuries are fairly minor, but often require treatment. Do you know the right way to get a sliver of copper out of your hand? If your helper suffers from an electrical shock when a drill cord goes bad, do you know what to do? Many people don't possess good first-aid skills.

Before we get too far into this chapter, there are a few main points to make. First, I'm not a medical doctor or any type of trained medical-care person. I've taken first-aid classes, but I'm certainly not an authority on medical issues. The suggestions in this chapter are for informational purposes only. This book is not a substitute for first-aid training offered by qualified professionals.

My intent here is to make you aware of some basic first-aid procedures that can make life on the job much easier. But it is important to understand that this book is not a first-aid manual. Hopefully, this chapter will show you the types of advantages you can gain from taking first-aid classes. Before you attempt first aid on anyone, including yourself, you should attend a structured, approved first-aid class. The information here is accurate, but you should not consider it complete. Take a little time to seek professional training in the art of first aid. You may never use what you learn, but the one time it is needed, you will be glad you made the effort to learn what to do. With this said, let's jump right into some tips on first aid.

OPEN WOUNDS

Open wounds are a common problem on construction sites. Many tools and materials used by workers can create open wounds. Here is what you should do if somebody suffers a cut:

- Stop the bleeding as soon as possible.
- Disinfect and protect the wound from contamination.

doi: 10.1016/B978-1-85617-549-4.00011-3

- Take necessary steps to avoid shock symptoms.
- Once the patient is stable, seek medical attention for severe cuts.

When a bad cut is encountered, the victim may slip into shock. A loss of consciousness could result from a loss of blood. Death from extreme bleeding is also a risk. As a first-aid provider, you must act quickly to reduce the risk of serious complications.

Bleeding

To stop bleeding, apply direct pressure on the wound. This may be as crude as clamping your hand over the wound, but a cleaner compression is preferable. Ideally, a sterile material should be placed over the wound and secured, normally with tape (even if it's duct tape). Whenever possible, wear rubber gloves to protect yourself from possible disease transfer if you are working on someone else. Thick gauze used as a pressure material can absorb blood and allow the clotting process to begin.

Bad wounds may bleed right through the compress material. If this happens, don't remove the blood-soaked material. Add a new layer of material over it. Keep pressure on the wound. If you are not prepared with a first-aid kit, you could substitute gauze and tape with strips cut from clothing that can be tied in place over the wound.

When you are dealing with a bleeding wound, it is usually best to elevate it. If you suspect a fractured or broken bone in the area of the wound, elevation may not be practical. When we talk about elevating a wound, it simply means to raise the wound above the level of the victim's heart. This helps the blood flow to slow down, due to gravity.

> Tourniquets get a lot of attention in movies, but they can do as much harm as good if not used properly. A tourniquet should only be used in a life-threatening situation.

Super serious bleeding

Super serious bleeding might not stop even after a compression bandage is applied and the wound is elevated. When this is the case, you must resort to putting pressure on the main artery that is producing the blood. Constricting an artery is not an alternative for the steps we have discussed previously.

Putting pressure on an artery is serious business. First, you must be able to locate the artery, and you should not keep the artery constricted any longer than necessary. You may have to apply pressure for a few minutes, release it, and then apply it again. It's important that you do not restrict the flow of blood in arteries for long periods of time. I hesitate to go into too much detail on this process, as I feel it is a method that you should be taught in a controlled, classroom situation. However, I will discuss the basics. Remember that the information here is not a substitute for professional training from qualified instructors.

Open arm wounds are controlled with the brachial artery. The location of this artery is in the area between the biceps and triceps, on the inside of the arm. It's about halfway between the armpit and the elbow. Pressure is created with the flat parts of your fingertips. Basically, you are holding the victim's wrist with one hand and closing off the artery with your other hand. Pressure exerted by your fingers pushes the artery against the arm bone and restricts blood flow. Again, don't attempt this type of first aid until you have been trained properly in the execution of the procedure.

Severe leg wounds may require the constriction of the femoral artery. This artery is located in the pelvic region. Normally, bleeding victims are placed on their backs for this procedure. The heel of a hand is placed on the artery to restrict blood flow. In some cases, fingertips are used to apply pressure. Again, you should seek professional training in these techniques.

When someone is impaled on an object, *do not remove the object.* Call professional help and leave the object in place. Removing it could result in more excessive bleeding.

Tourniquets

Tourniquets get a lot of attention in movies, but they can do as much harm as good if not used properly. A tourniquet should only be used in a life-threatening situation. When a tourniquet is applied, there is a risk of losing the limb to which the restriction is applied. This is obviously a serious decision and one that must be made only when all other means of stopping blood loss have been exhausted.

Unfortunately, concrete workers might run into a situation where a tourniquet is the only answer. For example, if a worker allowed a power saw to get out of control, a hand might be severed, or some other type of life-threatening injury could occur. This would be cause for the use of a tourniquet. Let's look at a few scenarios of when a tourniquet is used.

Tourniquets should be at least 2 inches wide. A tourniquet should be placed at a point that is above a wound, between the bleeding and the victim's heart. However, the binding should not encroach directly on the wound area. Tourniquets can be fashioned out of many materials. If you are using strips of cloth, wrap the cloth around the limb that is wounded and tie a knot in the material. Use a stick, screwdriver, or whatever else you can lay your hands on to tighten the binding.

Once you have made a commitment to apply a tourniquet, the wrapping should be removed only by a physician. It's a good idea to note the time that a tourniquet is applied, as this will help doctors later in assessing their options. As an extension of the tourniquet treatment, you will most likely have to treat the patient for shock.

Infection

Infection is always a concern with open wounds. When a wound is serious enough to require a compression bandage, don't attempt to clean the cut. Keep pressure on the wound to stop bleeding. In cases of severe wounds, be on the lookout for shock symptoms, and be prepared to treat them. Your primary concern with a serious open wound is to stop the bleeding and get professional medical help as soon as possible.

Lesser cuts, which are more common than deep ones, should be cleaned. Regular soap and water can be used to clean a wound before applying a bandage. Remember, we are talking about minor cuts and scrapes at this point. Flush the wound generously with clean water. A piece of sterile gauze can be used to pat the wound dry. Then a clean, dry bandage can be applied to protect the wound while in transport to a medical facility.

Fast facts

- Use direct pressure to stop bleeding.
- Wear rubber gloves to prevent direct contact with a victim's blood.
- When feasible, elevate the part of the body that is bleeding.
- Extremely serious bleeding can require you to put pressure on the artery supplying the blood to the wound area.
- Tourniquets can do more harm than good.
- Tourniquets should be at least 2 inches wide.
- Tourniquets should be placed above the bleeding wound, between the bleeding and the victim's heart.
- Tourniquets should not be applied directly on the wound area.
- Tourniquets should only be removed by trained medical professionals.
- If you apply a tourniquet, note the time that you apply the tourniquet.
- When a bleeding wound requires a compression bandage, don't attempt to clean the wound. Simply apply compression quickly.
- Watch victims with serious bleeding for symptoms of shock.
- Lesser bleeding wounds should be cleaned before being bandaged.

SPLINTERS AND SUCH

Splinters and other foreign objects often invade the skin of people on construction jobs. Getting these items out cleanly is best done by a doctor, but there are some on-the-job methods you might want to try. A magnifying glass and a pair of tweezers work well together when removing embedded objects, such as splinters and slivers of copper tubing. Ideally, the tweezers should be sterilized either over an open flame, such as the flame of your torch, or in boiling water.

Splinters and slivers that are submerged beneath the skin can often be lifted out with the tip of a sterilized needle. The use of a needle in conjunction with a pair of tweezers is very effective in the removal of most simple splinters. If you are dealing with something that has gone extremely deep into tissue, it is best to leave the object alone until a doctor can remove it.

EYE INJURIES

Eye injuries are very common on construction and remodeling jobs. Most of these injuries could be avoided if proper eye protection was worn, but far too many workers don't wear safety glasses and goggles. This sets the stage for eye irritations and injuries.

Before you attempt to help someone who is suffering from an eye injury, you should wash your hands thoroughly. I know this is not always possible on construction sites, but cleaning your hands is advantageous. In the meantime, keep the victim from rubbing the injured eye. Rubbing can make matters much worse.

Never attempt to remove a foreign object from someone's eye with the use of a rigid device, such as a toothpick. Cotton swabs that have been wetted can serve well as a magnet to remove some types of invasion objects. If the person you are helping has something embedded in an eye, get the person to a doctor as soon as possible. Don't attempt to remove the object yourself.

When you are investigating the cause of an eye injury, you should pull down the lower lid of the eye to determine if you can see the object causing the problem. A floating object, such as a piece of sawdust trapped between an eye and an eyelid, can be removed with a tissue, a damp cotton swab, or even a clean handkerchief. Don't allow dry cotton material to come into contact with an eye.

If looking under the lower lid doesn't reveal the source of discomfort, check under the upper lid. Clean water can be used to flush out many eye contaminants without much risk of damage to the eye. Objects that cannot be removed easily should be left alone until a physician can take over.

- Wash your hands, if possible, before treating eye injuries.
- Don't rub an eye wound.
- Don't attempt to remove embedded items from an eye.
- Clean water can be used to flush out some eye irritants.

SCALP INJURIES

Scalp injuries can be misleading. What looks like a serious wound can be a fairly minor cut. On the other hand, what appears to be only a cut can involve a fractured skull. If you or someone around you sustains a scalp injury, such as having a hammer fall on your head from an overhead worker, take it seriously. Don't attempt to clean the wound. Expect profuse bleeding.

If you don't suspect a skull fracture, raise the victim's head and shoulders to reduce bleeding. Try not to bend the neck. Put a sterile bandage over the wound, but don't apply excessive pressure. If there is a bone fracture, pressure could worsen the situation. Secure the bandage with gauze or some other material that you can wrap around it. Seek medical attention immediately.

FACIAL INJURIES

Facial injuries can occur on jobs. I've seen helpers let their right-angle drills get away from them, with the result being hard knocks to the face. On one occasion, I remember a tooth being lost, and split lips and tongues that have been bitten are common when a drill goes on a rampage.

Extremely bad facial injuries can cause a blockage of the victim's air passages. This, of course, is a very serious condition. It's critical that air passages be open at all times. If the person's mouth contains broken teeth or dentures, remove them. Be careful not to jar the individual's spine if you have reason to believe there may be injury to the back or neck.

Conscious victims should be positioned, when possible, so secretions from the mouth and nose will drain out. Shock is a potential concern in severe facial injuries. For most on-the-job injuries, plumbers should be treated for comfort and sent for medical attention.

NOSEBLEEDS

Nosebleeds are not usually difficult to treat. Typically, pressure applied to the side of the nose where bleeding is occurring will stop the flow of blood. Applying cold compresses can also help. If external pressure does not stop the bleeding, use a small, clean pad of gauze to create a dam on the inside of the nose. Then, apply pressure on the outside of the nose. This will almost always work. If it doesn't, get to a doctor.

Falls can result in all sorts of hidden injuries. These could include concussion or a broken back. Don't attempt to move fall victims. Call in the professionals and keep the victim stable.

BACK INJURIES

There is really only one thing you need to know about back injuries: *Don't move the injured party.* Call for professional help, and see that the victim remains still until help arrives. Moving someone who has suffered a back injury can be very risky. Don't do it unless there is a life-threatening cause for your action, such as a person trapped in a fire or some other type of deadly situation.

LEGS AND FEET

Legs and feet sometimes become injured on job sites. The worst case of this type that I can remember was when a plumber knocked a pot of molten lead over on his foot. (It still sends shivers up my spine just thinking about it.) Anyway, when someone suffers a minor foot or leg injury, you should clean and cover the wound. Bandages should be supportive without being constrictive. The extremity should be elevated above the victim's heart level when possible. Prohibit the person from walking. Remove boots and socks so you can keep an eye on the person's toes. If the toes begin to swell or turn blue, loosen the supportive bandages.

There are certain signs of shock that you can look for. If a person's skin turns pale or blue and is cold to the touch, those are signs of shock. Skin that becomes moist and clammy can indicate that shock is present. General weakness is also a sign of shock.

Blisters

Blisters may not seem like much of an emergency, but they can sure take the steam out of a hard worker. In most cases, blisters can be covered with a heavy gauze pad to reduce pain. It is generally recommended to leave blisters unbroken. When a blister breaks, the area should be cleaned and treated as an open wound. Some blisters tend to be more serious than others. For example, blisters in the palm of a hand or on the sole of a foot should be looked at by a doctor.

HAND INJURIES

Hand injuries are common in the trades. Little cuts are the most frequent complaint. Serious hand injuries should be elevated. This tends to reduce swelling. You should not try to clean severe hand injuries. Use a pressure bandage to control bleeding. If the cut is on the palm of a hand, a roll of gauze can be squeezed by the victim to slow the flow of blood. Pressure should stop the bleeding, but if it doesn't, seek medical assistance. As with all injuries, use common sense on whether professional attention is required after first aid is applied.

SHOCK

Shock is a condition that can be life threatening even when the injury responsible for a person going into shock is not otherwise fatal. We are talking about traumatic shock, not electrical shock. Many factors can lead to a person going into shock. A serious injury is a common cause, but many other causes exist. There are certain signs of shock that you can look for.

If a person's skin turns pale or blue and is cold to the touch, that indicates shock. Skin that becomes moist and clammy can indicate shock is present. General weakness is also a sign of shock. When a person is going into shock, the individual's pulse is likely to exceed 100 beats per minute. Breathing is usually increased, but it may be shallow, deep, or irregular. Chest injuries usually result in shallow breathing. Victims who have lost blood may be thrashing about as they enter into shock. Vomiting and nausea can also signal shock.

As a person slips into deeper shock, the individual may become unresponsive. Look at the eyes; they may be dilated. Blood pressure can drop, and in time, the victim will lose consciousness. Body temperature will fall, and death will be likely if treatment is not rendered.

There are three main goals when treating someone for shock: (1) get the person's blood circulating well, (2) make sure an adequate supply of oxygen is available to the individual, and (3) maintain the person's body temperature.

When you have to treat a person for shock, you should keep the victim lying down. Cover the individual so the loss of body heat will be minimal. Get medical help as soon as possible. The reason it's best to keep a person lying down is that the individual's blood should circulate better. Remember, if you suspect back or neck injuries, don't move the person.

People who are unconscious should be placed on one side so fluids can run out of the mouth and nose. It's also important to make sure that air passages are open. A person with a head injury may be laid out flat or propped up, but the head should not be lower than the rest of the body. It is sometimes advantageous to elevate a person's feet when he or she is in shock. If there is any difficulty in breathing, however, or if pain increases when the feet are raised, lower them.

Body temperature is a big concern with shock patients. You want to overcome or avoid chilling. However, don't attempt to add additional heat to the surface of the person's body with artificial means. This can be damaging. Use only blankets, clothes, and other similar items to regain and maintain body temperature.

Avoid the temptation to offer the victim fluids, unless medical care is not going to be available for a long time. Avoid fluids completely if the person is unconscious or is subject to vomiting. Under most job-site conditions, fluids should not be administered.

Checklist of shock symptoms

- ✓ Skin that is pale, blue, or cold to the touch
- ✓ Skin that is moist and clammy
- ✓ General weakness
- ✓ Pulse rate in excess of 100 beats per minute
- ✓ Increased breathing
- ✓ Shallow breathing
- ✓ Thrashing
- ✓ Vomiting and nausea
- ✓ Unresponsive action
- ✓ Widely dilated eyes
- ✓ A drop in blood pressure

BURNS

Burns are not typically common among concrete installers, but they can occur in the workplace. There are three types of burns that you may have to deal with. First-degree burns are the least serious. These burns typically come from overexposure to the sun, which construction workers often suffer from; quick contact with a hot object, like the tip of a torch; and scalding water, which could be the case when working with a boiler or water heater.

Second-degree burns are more serious. They can come from a deep sunburn or from contact with hot liquids and flames. A person who is affected by a second-degree burn may have a red or mottled appearance, blisters, and a wet appearance of the skin within the burn area. This wet look is due to a loss of plasma through the damaged layers of skin.

Third-degree burns are the most serious. They can be caused by contact with open flames, hot objects, or immersion in very hot water. Electrical injuries can also result in third-degree burns. This type of burn can look similar to a second-degree burn, but the difference will be the loss of all layers of skin.

Treatment

Treatment for most job-related burns can be administered on the job site and will not require hospitalization. First-degree burns should be washed with or submerged in cold water. A dry dressing can be applied if necessary. These burns are not too serious. Eliminating pain is the primary goal with first-degree burns.

Second-degree burns should be immersed in cold (*not* ice) water. The soaking should continue for at least one hour and up to two hours. After soaking, the wound should be layered with clean cloths that have been dipped in ice water and wrung out. Then the wound should be dried by blotting, not rubbing. A dry, sterile gauze should then be applied. Don't break open any blisters. It is also not advisable to use ointments and sprays on severe burns. Burned arms and legs should be elevated, and medical attention should be acquired.

Bad burns, the third-degree type, need quick medical attention. First, don't remove a burn victim's clothing, since skin might come off with it. A thick, sterile dressing can be applied to the burn area. Personally, I would avoid this if possible. A dressing might stick to the mutilated skin and cause additional skin loss when the dressing is removed. When hands are burned, keep them elevated above the victim's heart. The same goes for feet and legs. You should not soak a third-degree burn in cold water. It could induce more shock symptoms. Don't use ointments, sprays, or other types of treatments. Get the burn victim to competent medical care as soon as possible.

> You owe it to yourself, your family, and the people you work with to learn first-aid techniques. This can be done best by attending formal classes in your area. Most towns and cities offer first-aid classes on a regular basis. I strongly suggest that you enroll in one.

HEAT-RELATED PROBLEMS

Heat-related problems can include heatstroke and heat exhaustion. Cramps are also possible when working in hot weather. Anybody who doesn't think heatstroke is serious is wrong. Heatstroke can be life threatening. People affected by heatstroke can develop body temperatures in excess of 106°F. Their skin is likely to be hot, red, and dry. You might think sweating would take place, but it doesn't. Pulse is rapid and strong, and victims can slip into unconsciousness.

If you are dealing with heatstroke, you need to lower the person's body temperature quickly. There is a risk, however, of cooling the body too quickly once the victim's temperature is below 102°F. You can lower body temperature with rubbing alcohol, cold packs, cold water on cloths, or in a bathtub of cold water. Avoid the use of ice in the cooling process. Fans and air-conditioned space can be used to achieve your cooling goals. Get the body temperature down to at least 102°F and then go for medical help.

Cramps

Cramps are not uncommon among workers during hot spells. A simple massage can be all it takes to cure this problem. Saltwater solutions are another way to control cramps. Mix one teaspoon of salt in 8 ounces of water, and have the victim drink half a glass about every 15 minutes.

Exhaustion

Heat exhaustion is more common than heatstroke. A person affected by heat exhaustion is likely to maintain a fairly normal body temperature, but the skin may be pale and clammy. Sweating may be very noticeable, and the individual will probably complain of being tired and weak. Headaches, cramps, and nausea may accompany the symptoms. In some cases, fainting might occur.

The saltwater treatment described for cramps will normally work with heat exhaustion. Victims should lie down and elevate their feet about a foot off the floor or bed. Clothing should be loosened, and cool, wet cloths can be used to add comfort. If vomiting occurs, get the person to a hospital for intravenous fluids.

We could talk about first aid in much more length and detail, but in a book like this, any discussion of medical procedures must be limited. You owe it to yourself, your family, and the people you work with to learn first-aid techniques. This can be done best by attending formal classes in your area. Most towns and cities offer first-aid classes on a regular basis. Until you have some hands-on experience in a classroom and gain the depth of knowledge needed, you are not prepared for emergencies. Don't get caught short. Prepare now for the emergency that hopefully will never happen.

CHAPTER

Working with the concrete code 12

Using the code in the real world is a little different from reading and understanding a codebook. In theory, the book should hold all the answers, and you should be able to apply them without incurring any unexpected problems. But reality is not always so simple. Putting the concrete code to practical use doesn't have to be difficult, but it can be.

There are two extremes to applying code requirements on a job. On one side, you have situations where the code enforcement is lax. Then there are the times when the code enforcement is extremely strict. Most jobs run somewhere between these two extremes. Whether the job you are working on is lax or strict, you could have problems dealing with code requirements.

LAX JOBS

There are some workers who welcome lax jobs. These people enjoy not having their work scrutinized too closely. I guess everyone might enjoy an easy job, but there are risks to lax jobs. In my opinion, professional concrete workers should perform professional services, even if they can get away with less-than-credible work. This should be an ethical commitment. But there are workers who will cut corners in jurisdictions where they know they can get away with it. This is not fair to the customer, and it can put the installer at risk.

Inspectors provide a form of protection for both installers and contractors. Cutting corners can put you at great risk for a lawsuit. Leaving the financial side of such a suit out of the picture, try to imagine how you would deal with the guilt of personal injury to people as a result of your deviation from the code.

How many times have you installed new work without a permit? Did you know that a permit and inspection were required? Many workers have skipped the permit and inspection element of a job at one time or another. But they probably never thought of the risk. What would happen if the concrete installation failed? Would

doi: 10.1016/B978-1-85617-549-4.00012-5

people be hurt? Could anyone be killed? What is the most extreme risk that you could be subject to? Do you really want to lose sleep at night over cutting corners on permits? It is not worth it. When a permit is required, get one.

> Your liability insurance may not be willing to cover any claim made for work that you did without following code requirements. If you had done the job by the book, you would have some foundation for defending yourself. Additionally, you would have an approved inspection certificate that would validate the fact that you did the work properly and within code requirements. This could go a long way in a lawsuit.

The small amount of money you save by cheating on permits and inspection could come back as a massive lawsuit and wipe out your career or your business. When you work in an area prone to lax inspections, you can be at risk. It is in your best interest to do all work to code requirements and to have that work inspected and approved.

STRICT CODE ENFORCEMENT

Strict code enforcement can be a blessing. It helps to take you off the hook if something goes wrong down the road. On the other hand, it can be a real pain when you are being ripped apart on a simple job. Again, you have to use common sense in your response to such situations.

Is it fair if an inspector fails your job because you are missing one anchor? Technically, yes. In reality, especially if the code officer knows that you are reputable and professional, what would it hurt to pass the job with a notice for you to install the missing anchor?

SAFETY

The concrete code exists for a reason. Safety is what the code is focused on, and this is a valid reason for code regulations. Most installers have felt from one time to another that code regulations are too strict or not needed for certain elements of the business. While this may be true on occasion, the basic foundation of the code is a sound one, and it should be observed. There are elements of code regulations that I feel are too detailed for practical purposes, but I always attempt to comply with code requirements.

FEES

Both contractors and homeowners complain about the fees charged for permits. These fees pay for the administration of the code and the protection of consumers.

Indirectly, contractors receive protection from inspections. When you feel taken advantage of due to costly fees charged by a code enforcement office, consider the benefits that our society is receiving. I am not trying to be a proponent of code fees, but I do believe that people in the trades should look at the fees in a reasonable fashion.

KNOW YOUR INSPECTORS

To make your life easier, get to know your inspectors. Communication is a major portion of any good deal. Knowing the inspectors in your area and what they expect will make jobs run more smoothly. Most inspectors are accessible and helpful as long as you don't approach them with a bad attitude.

> When doing work in a new location for the first time, it is often helpful to meet with the local code officers to introduce yourself and find out what they expect from you.

LOCAL JURISDICTIONS

Local jurisdictions adopt a code and have the right to amend it for local requirements. In simple terms, this can mean that a local code may vary from the base code that is adopted. Even if you know the primary code by heart, you might have a conflict with a local jurisdiction. A visit with your local code officers will clear up such differences before they become a problem on a job.

COMMON SENSE

Common sense goes a long way in the installation of concrete systems. It is not uncommon for a mixture of common sense and code requirements to be used to achieve a successful installation. Code regulations must be followed, but they are not always as rigid as they may appear to be. Don't be afraid to consult with inspectors to arrive at a reasonable solution to difficult problems.

In closing, using the code in the real world is largely a matter of proper communication between contractors and code officers. This is an important fact to keep in mind. Learn the code and use it properly. If you feel a need to deviate from it, talk with a code officer and arrive at a solution that is acceptable to all parties affected. If you keep a good attitude and don't scrap the code because you think you know better, you should do well in your trade.

CHAPTER

Test your code knowledge

13

How much do you know about the code pertaining to concrete work? Are you someone who feels that knowing the code should be someone else's job? If you are going to make a career in concrete, you must understand the code. This chapter gives you a broad-brush look at the types of requirements specified in the code. If you will take the time to test your existing knowledge, you will gain insight into how much effort you should put forth to learn various elements of the code.

What follows are quizzes on concrete code facts. Some are in true-false format. Others are in a multiple-choice form. Answers are found at the end of the chapter. I am going to work though the codebook and pose questions for you to answer. Some will be simple. Many of the questions will pertain to day-to-day concrete work. Some of the questions will be more removed from daily life.

No one expects you to memorize the concrete code. It is sufficient to be able to look up answers to your questions in the codebook when you are on the job. However, this can make customers a bit nervous about your knowledge and ability. At the very least, you need to know what questions to ask yourself when doing a job. Let's start with some true-false questions and see how you do.

TRUE OR FALSE QUIZ

1. It is necessary for all equipment for mixing and transporting concrete to be clean.
   True False
2. Areas to be filled with concrete must be free of ice and debris.
   True False
3. Concrete forms must be protected from being coated at all times.
   True False
4. Masonry filler units are not allowed to come into contact with concrete.
   True False

doi: 10.1016/B978-1-85617-549-4.00013-7

5. Reinforcement must be thoroughly clean of ice.
   True False
6. Reinforcement must be thoroughly clean of deleterious coatings.
   True False
7. Concrete must be deposited within 50 feet of its final destination to avoid segregation during handling.
   True False
8. Concreting is to be done at such a rate that concrete is at all times plastic and flows readily into spaces between reinforcement.
   True False
9. Preferred curing conditions and time for concrete requires the curing to be done at temperatures above 62°F and in moist conditions for at least 9 days.
   True False
10. Curing concrete with high-pressure steam is prohibited.
    True False
11. When concrete is cured with an accelerated process, the compressive strength of the concrete at the load stage must be considered at least equal to the required design strength at that load stage.
    True False
12. Many installers are aware of the need for special provisions when concreting in cold temperatures. Hot-temperature concreting offers its own challenges and requires special consideration for water evaluation and related risks.
    True False
13. When repairing existing concrete, forms for concreting must be installed and supported to avoid any damage to existing concrete.
    True False
14. The rate and method of placing concrete must be factored into the design of formwork for concreting.
    True False
15. Conduits and pipes, with their fittings, embedded within a column shall not displace more than 7.5 percent of the area of cross section on which strength is calculated or which is required for fire protection.
    True False
16. Pipes embedded in concrete shall not hold or convey any liquid, gas, or vapor, except water not exceeding 90°F or 50 PSI pressure until design strength is obtained.
    True False
17. The surface of concrete construction joints must be cleaned without removing any existing laitance.
    True False

18. Construction joints must be wetted and standing water removed prior to placing new concrete.
True False

19. Construction joints in girders shall be offset a minimum distance of two times the width of intersecting beams.
True False

20. Construction joints in floors shall be located within the middle third of spans of slabs, beams, and girders.
True False

21. All bending of reinforcement material is to be done while the material is hot.
True False

22. When dealing with bundled bars, the groups of parallel reinforcing bars bundled in contact to act as a unit shall be limited to six in one bundle.
True False

23. Bars larger than Number 11 shall not be bundled in beams.
True False

24. Bundled bars shall be enclosed within stirrups or ties.
True False

25. In walls and slabs other than concrete joist construction, primary flexural reinforcement shall not be spaced farther apart than three times the wall or slab thickness nor farther apart than 18 inches.
True False

How do you think you have done so far? Do the correct answers jump out at you, or do you have to scratch your head from time to time to come up with an answer? Are you curious about your score? Well, take a moment and check your answers against the correct answers at the end of the chapter. Take a little break, and then we will do the multiple-choice section of the test.

How did you do? How much do you *really* know about the concrete code? Are you good to go, or do you need to brush up on the code? Are you ready to try the multiple-choice questions? Let's do it.

MULTIPLE-CHOICE QUESTIONS

1. Generally speaking, bundled bars must be covered with a minimum amount of concrete equal to the diameter of the bundle. The maximum cover under general conditions is:
A. 2 inches B. 3 inches C. 4 inches D. 6 inches

2. When bundled bars exist in concrete that is going to be cast against and permanently exposed to earth, the minimum amount of coverage allowed is:
   A. 2 inches B. 3 inches C. 4 inches D. 6 inches

3. When working with offset bars, the slope of the inclined portion of an offset bar with the axis of the column shall not exceed:
   A. 1 in 6 B. 3 in 6 C. 4 in 6 D. 5 in 6

4. Horizontal support at offset bends shall be provided by lateral ties, spirals, or parts of the floor construction. Horizontal support provided shall be designed to resist _______ times the horizontal component of the computed force in the inclined portion of an offset bar.
   A. 1 B. 1.5 C. 2.25 D. 3

5. Lateral ties or spirals, if used, shall be placed not more than _______ inches from points of bend.
   A. 2 B. 4 C. 6 D. 8

6. Offset bars shall be bent _______ placement in forms.
   A. before B. after C. during D. in

7. Enclosure at connections shall consist of external concrete or internal closed ties, _______, or stirrups.
   A. racks B. leverage C. spirals D. concaves

8. For cast-in-place construction, size of spirals shall not be less than _______ inch in diameter.
   A. 18 B. 1/4 C. 1/2 D. 3/8

9. Clear spacing between spirals shall not exceed _______ inches.
   A. 2 B. 3 C. 6 D. 9

10. Vertical spacing of ties shall not exceed _______ longitudinal bar diameters.
    A. 4 B. 10 C. 12 D. 16

11. Vertical spacing of ties shall not exceed _______ tie bar or wire diameters, or the least dimension of the compression member.
    A. 12 B. 24 C. 48 D. 62

12. Shrinkage and temperature reinforcement shall be spaced not farther apart than five times the slab thickness nor farther apart than _______ inches.
    A. 4 B. 6 C. 12 D. 18

13. The spacing of tendons must not exceed _______ feet.
    A. 3 B. 6 C. 9 D. 12

14. When dealing with joist construction, clear spacing between ribs shall not exceed _______ inches.
    A. 16 B. 22 C. 24 D. 30

15. Slab thickness in joist construction must not be less than 1/12 the clear distance between ribs nor less than _____ inches.
A. 2 B. 6 C. 24 D. 30

16. Design yield strength of structural steel core shall be the specified minimum yield strength for the grade of structural steel used but not to exceed _________ PSI.
A. 100 B. 10,000 C. 50,000 D. 75,000

17. A shearhead shall not be deeper than _____ times the web thickness of the steel shape.
A. 10 B. 20 C. 50 D. 70

18. Mechanical and welded splices ________ be permitted.
A. shall B. shall not C. may D. are never to

19. Walls more than _______ inches thick, except basement walls, shall have reinforcement for each direction placed in two layers parallel with faces of the wall.
A. 4 B. 6 C. 8 D. 10

20. Thickness of nonbearing walls shall not be less than ______ inches nor less than 1/30 the least distance between members that provide lateral support.
A. 4 B. 6 C. 8 D. 10

21. The depth of footing above bottom reinforcement shall not be less than _______ inches for footings on soil.
A. 4 B. 6 C. 8 D. 10

22. The depth of footing above bottom reinforcement shall not be less than _______ inches for footings on piles.
A. 6 B. 8 C. 10 D. 12

23. For structural integrity, ties around the perimeter of each floor and roof within 4 feet of the edge shall provide a nominal strength in tension not less than ________ pounds.
A. 10,000 B. 12,000 C. 16,000 D. 24,000

24. The use of an entire composite member or portions thereof for resisting shear and moment shall ________.
A. be permitted B. not be permitted C. exist D. be discouraged

25. Three-dimensional effects shall be considered in design and analyzed using three-dimensional procedures or approximated by considering the summation of effects of _____ orthogonal planes.
A. 2 B. 3 C.4 D. 6

This amounts to 50 questions and answers that pertain to the concrete code. Check your answers to determine your natural knowledge of the code. Remember, the code is a big part of your career in concrete. If you don't know or understand the code requirements well, spend some time studying the code. It will be a good investment of your time.

ANSWERS TO TRUE-FALSE QUESTIONS

1. T
2. T
3. F
4. F
5. T
6. T
7. F
8. T
9. F
10. F
11. T
12. T
13. T
14. T
15. T
16. T
17. F
18. T
19. T
20. T
21. T
22. F
23. T
24. T
25. T

ANSWERS TO MULTIPLE-CHOICE QUESTIONS

1. A
2. B
3. A
4. B
5. C
6. A
7. C
8. D
9. C
10. D
11. C
12. D
13. B
14. D
15. A
16. C
17. D
18. A
19. D
20. A
21. B
22. D
23. C
24. A
25. A

CHAPTER

Construction code highlights

14

We don't have enough room in this book to talk about the entire concrete code. The codebook itself is about 430 pages long. This, however, does not prevent us from hitting the highlights of the code. What you will find here are tips, tidbits, and commentary on many elements of the concrete code. This chapter is not meant to replace your codebook, but it can give you a jump start on some of the more commonly used code requirements. With that said, let's get started.

CONCRETE SELECTION

Proportions of concrete must be selected. Before this can be done properly, you must confirm that your desire for the selection is in compliance with the concrete code. How will the workability and consistency of the concrete be? Will you be able to work the mixture into forms and around reinforcement under conditions of placement to be employed without segregation or excessive bleeding? The mixture must be workable and have a quality consistency.

> The code provides tables for you to use as reference points in establishing various conditions and requirements.

Before a concrete material can be used, it must be approved to resistance to special exposures. The mixture must also be in conformance with strength test requirements. When different materials are planned for use in different portions of proposed work, the combination of the materials being used together must be considered and determined effective.

doi: 10.1016/B978-1-85617-549-4.00014-9

COMPRESSIVE STRENGTH

Test records used to demonstrate proposed concrete proportions are required to represent materials and conditions similar to those expected to be used. When there are changes in materials, conditions, and proportions within the test records, they must not be more restricted than those for proposed work.

Tests for compressive strength require not less than 30 test records. Ten consecutive tests are acceptable when the test records cover a period of time that is not less than 45 days. When an acceptable record of field test results is not available, there are alternative methods to compensate for the lack of test records.

Trial mixtures with the proportions and consistencies required for proposed work require the use of a minimum of three different water–cementitious materials contents that will provide a range of strengths encompassing compressive strength.

The slump of a trial mixture must be designed not to exceed ± 0.75 inches of the maximum permitted and for air-entrained concrete, within ± 0.5 percent of the maximum allowable air content.

When cylinder testing is done, the test must be done for 28 days or at the test age designated for the proper calculation of compressive strength. Results from these tests can plot a curve to show the relationship between the water–cementitious materials ratio or cementitous materials content and compressive strength at a set test age.

FIELD-CURED SPECIMENS

Test results from field-cured specimens tested in cylinders may be required for the inspection of code officers. All field-cured test cylinders are required to be molded at the same time and from the same samples as test cylinders that have been tested under laboratory conditions.

The requirements for protecting and curing concrete must be improved when strength of field-cured cylinders at the test age designated for determination of compressive strength is less than 85 percent of that found in companion laboratory-cured cylinders. This rule is not required when the field-cured strength exceeds the compressive strength requirements by more than 500 psi.

Should test results indicate low-strength concrete and show that the load-carrying capacity is heavily reduced, test cores drilled for the area in question are allowed. Core samples must be wiped dry prior to transport. The core samples are to be placed in watertight bags or containers immediately after drilling. Core samples must not be tested sooner than 48 hours after being taken or more than 7 days after coring. The only exception to this is if a registered design professional delivers an acceptable alternative.

What makes concrete in an area represented by coring acceptable? There must be three cores that are equal to at least 85 percent of the compressive strength required. It is also necessary that no more than one core sample is less than 75 percent of compressive strength. It is acceptable to take core samples from locations represented by erratic core strength.

PUTTING CONCRETE IN PLACE

There are rules and regulations that apply to the installation of concrete. For example, all equipment used to mix or transport concrete must be clean. Any ice present on the work section must be removed. Concrete forms are required to be coated prior to the pouring process. If you will be using masonry filler units that will be in contact with concrete, the fillers must be drenched with water. All reinforcement material must be clean of ice and deleterious coatings. Water in the pour site must be removed. Any loose material in the pour area must be removed prior to concrete installation.

MIXING CONCRETE

When mixing ready-mix concrete, the concrete must be mixed until there is a uniform distribution of materials. The full contents of the mixer must be emptied before the mixer is recharged.

Conditions change when concrete is mixed on a job site. A batch mixer of an approved type must be used for the mixing process. The operator of the mixer must know and operate the mixer at a speed recommended by the manufacturer of the mixer. Once all materials are in the drum of the mixer, the mixing process shall continue for a minimum of 90 seconds.

Detailed records must be kept of the mixing and placement procedures for job-mixed concrete. These records must include the following:

- Number of batches of concrete used
- Proportions of materials used
- Approximate location of final deposit in structure
- Time of mixing and placing
- Date of mixing and placing

Concrete must be conveyed in a manner to prevent separation or loss of materials.

PLACING CONCRETE

When placing concrete, place it as close as practical to its final position. This helps to avoid segregation due to rehandling or flowing. All concrete being placed must be plastic and must flow readily into spaces between reinforcements. Any bad concrete mix, such as a mixture that has begun to harden, must be disposed of properly and not used in the construction process.

Once concreting begins, it must continue continuously until a suitable stopping point is reached. Top surfaces of vertically formed lifts must be generally level. Concrete must be thoroughly consolidated by suitable means during placement and must be completely worked around reinforcement and embedded fixtures and into the corners of forms.

CURING

The curing of most concrete should be done at an air temperature of at least 50°F in moist conditions. This curing process should continue for a minimum of seven days. High-early-strength concrete can be cured in just three days under the same curing conditions.

High-pressure steam can be used to accelerate the curing process. Steam at atmospheric pressure, heat, and moisture can be used for curing concrete quickly. These circumstances can also be used to accelerate strength gain. When these techniques are used, the concrete must be equal to required design strength at the load stage.

Concrete curing in cold weather requires equipment to be available in order to keep materials from freezing. Concrete materials and all reinforcement, forms, fillers, and ground that will come into contact with concrete must be free of frost and ice. No frozen materials may be used in concreting.

When the weather is particularly hot, newly poured concrete must be protected against water evaporation. This can be done by dampening the concrete as needed. Tenting may be required to prevent excessive temperatures from affecting the curing of concrete.

FORM DESIGN

Form design results in a form that is made to create an encasement for concrete that will hold it in place during the curing process for the shapes and sizes desired. All forms must be built to prevent leakage. Appropriate bracing and restraints are required to keep forms in their proper placement. When you are rehabbing existing concrete, the forms for new concrete must be constructed in a way that will not harm the existing concrete.

A few factors come into play with form design. For instance, all forms must be built with consideration of the rate and method in which concrete will be placed in the form. Loads must be considered. This includes vertical, horizontal, and impact loads. Forms used for prestressed concrete members are to be designed and constructed to permit movement of the member without damage during application of prestressing force.

FORM REMOVAL

Safety and serviceability must not be compromised during the removal of forms. Concrete contained in a form must be cured to a strength suitable for exposure prior to the removal of any form work from the concrete.

A procedure must be developed that will show the process and schedule for the removal of shores and installation of reshores and for calculating the loads transferred to a structure during the process. Code officers may require documentation of the plans and procedures for form removal and shoring.

Form supports for prestressed concrete members are not to be removed before the members can support their dead load. Structural analysis is required to determine when concrete is strong enough to carry its required loads.

EMBEDDED ITEMS

Conduits, pipes, sleeves, and similar items may be embedded in concrete. Aluminum conduits must not be embedded in concrete unless the conduit is effectively coated or covered to prevent aluminum-concrete reaction or electrolytic action between aluminum and steel. When a conduit or pipe penetrates a wall or beam, the installation must be done in a manner to avoid impairing the strength of construction.

Columns may be used to conceal embedded items. However, the items must not displace more than 4 percent of the area of cross section on which strength is calculated or that is required for fire protection.

> Commonly embedded items must not have an outside dimension that is more than one-third the overall thickness of the slab, wall, or beam in which the embedding occurs.

Conduits and similar embedded items are not to be installed closer than three diameters or widths on center. Embedded items must not be exposed to rusting or other deterioration. Pipes used under concrete are to be uncoated or galvanized iron or steel not thinner than standard Schedule 40 steel pipe. The inside diameters of pipes must not exceed two inches. All embedded items must be designed to resist effects of the material, pressure, and temperature to which the materials will be subjected.

Until concrete reaches its design strength, embedded conduits are not allowed to convey liquid, gas, or vapor. One exception to this is water that does not exceed 90°F. The pressure of the water being conveyed must not exceed 50 psi.

Piping in solid slabs must be placed between the top and bottom reinforcement, unless the piping is to be used for radiant heat or snow melting. Concrete cover over embedded items must not be less than 1.5 inches when concrete is exposed to weather or the earth. When concrete is not exposed to exterior conditions, the minimum cover required is three-quarters of an inch.

Reinforcement with an area not less than 0.002 times area of concrete section shall be provided normal to piping.

CONSTRUCTION JOINTS

The surface of all construction joints must be cleaned, and all laitance must be removed. Before new concrete can be placed, all construction joints must be wetted, and all standing water shall be removed. The strength of a structure must not be impaired by construction joints. All construction joints should be located within the middle third of spans of slabs, beams, and girders. Vertical support members that are still plastic must not be used to support beams, girders, or slabs. Except when shown otherwise in design drawings or specifications, beams, girders, haunches, drop panels, and capitals are to be placed monolithically as part of a slab system.

Reinforcements that are partially embedded in concrete are not allowed to be bent in the field. The only exception to this is if the bending is shown on design drawings or is permitted by an engineer.

REINFORCEMENT

Reinforcement material must not be covered with mud, oil, ice, or any other element that will decrease the reinforcement's ability to bond. With the exception of prestressing steel, steel reinforcement with rust or mill scale or both can be determined to be satisfactory for use. If the reinforcement has the proper minimum dimensions and can be cleaned with a hand brush, the material should be acceptable. A light coating of rust is acceptable on prestressing steel.

Reinforcement, including tendons, and posttensioning ducts shall be accurately placed and adequately supported before concrete is placed.

Minimum clear spacing between parallel bars in a layer is not allowed to be less than 1 inch. Reinforcement bars placed in layers must be placed directly above any bars below with a minimum clear distance of 1 inch. Spirally reinforced or tied reinforced compression members require a minimum clear distance of 1.5 inches between reinforcements.

> Welding of crossing bars shall not be permitted for assembly of reinforcement unless authorized by an engineer.

When bundled bars are used as a reinforcement, they make a unit, and the unit shall not contain more than four bars. Stirrups and ties are used to house bundled bars. No bar larger than a Number 11 is to be used in a bundle that will be used in a beam.

> Bundling or posttensioning ducts shall be permitted if it is shown that concrete can be satisfactorily placed and if provision is made to prevent the prestressing steel, when tensioned, from breaking through the duct.

COLUMN REINFORCEMENT

Column reinforcement often includes the use of offset bars. The offset bent longitudinal bars must meet minimum requirements for code compliance. The slope of an inclined portion of an offset bar with axis of column is not allowed to exceed 1 in 6. The portions of bars above and below an offset are required to be parallel to axis of column. Offset bars that require bending must be bent prior to placing the bars in a concrete form.

Offset bends shall be provided horizontal support by lateral ties, spirals, or parts of the floor construction. The support must be designed to resist 1.5 times the horizontal component of the computed force in the inclined portion of an offset bar. When lateral ties or spirals are used for support, they must be placed no more than 6 inches from the points of bend.

> Longitudinal bars must not be offset bent where a column face is offset by 3 inches or more.

Ends of structural steel cores are to be accurately finished. They must bear at end-bearing splices, with positive provision for alignment of one core above the other in concentric contact. End-bearing splices are considered effective when they do not transfer more than 50 percent of the total compressive stress in the steel core. Transfer of stress between column base and footing shall be designed in compliance with the concrete code.

CONNECTIONS

In connections of principal framing elements, like beams, an enclosure shall be provided for splices of continuing reinforcement and for anchorage of reinforcement terminating in such connections. External concrete, internal closed ties, spirals, or stirrups can be used for the enclosure at connections.

SPIRALS

Spirals are to consist of evenly spaced, continuous bar or wire of certain size and assembled to permit handling and placing without distortion from designed dimensions. Spirals for cast-in-place construction require a minimum diameter of ⅜ inch. The clear spacing between spirals is not to exceed 3 inches. The minimum clear spacing between spirals is not to be less than 1 inch. The anchorage of spiral reinforcement must be provided by 1.5 extra turns of spiral bar or wire at each end of a spiral unit. Spirals are required to extend from the top of a footing or slab in any story to the level of the lowest horizontal reinforcement in members that are supported above.

Spirals must be held in place and be true to line.

TIES

Beams and brackets are not always framed into all sides of a column. When this is the case, ties are required. The ties are to extend above termination of the spiral to the bottom of the slab or drop panel. Columns that have capitals require spirals that extend to a level at which the diameter or width of the capital is twice that of the column.

Nonprestressed bars are to be enclosed by lateral ties at least Number 3 in size, for longitudinal bars Number 10 or smaller, and at least Number 4 in size for Numbers 11, 14, and 18 and bundled longitudinal bars. Deformed wire or welded wire reinforcement, or an equivalent area, shall be permitted.

The spacing between vertical ties is not to exceed 16 longitudinal bar diameters, 48 tie bar or wire diameters, or least dimension of the compression member.

Ties require a specific arrangement. All corner and alternate longitudinal bars require lateral support provided by the corner of a tie with an included angle of not more than 135 degrees, and no bar is allowed to be farther than 6 inches on each side along the tie from such a laterally supported bar. Longitudinal bars located around the perimeter of a circle may use a complete circular tie.

Ties are to be located vertically not more than one-half a tie spacing above the top of a footing or slab. Spacing must be not more than one-half a tie spacing below the lowest horizontal reinforcement in slab or drop panel.

Beams and brackets that are framed from four different directions into a column with termination of ties not more than 3 inches below lowest reinforcement in shallowest of such beams or brackets shall be permitted.

Lateral reinforcement for flexural framing members subject to stress reversals or to torsion at supports, shall consist of closed ties, closed stirrups, or spirals extending around the flexural reinforcement.

SHRINKAGE

Reinforcements for shrinkage are to be spaced no more than five times the slab thickness or no farther apart than 18 inches. Tendons must be proportioned so there is at least an average compressive stress of 100 psi on gross concrete areas using effective prestress after losses. The maximum allowable distance for spacing tendons is six feet.

STRUCTURAL INTEGRITY REQUIREMENTS

Cast-in-place construction requires certain minimum requirements. For example, beams along the perimeter of a structure require continuous reinforcement. At least two bars are required. Support shall consist of at least one-sixth of the tension reinforcement required for negative movement at the support. You must provide at least one-quarter of the tension reinforcement required for positive movement at midspan. When splices are needed to provide continuity, the top reinforcement is to be spliced at or near midspan. Bottom reinforcement must be spliced at or near the support. Class "A" tension splices are normally used.

Precast construction calls for the use of tension ties. They are to be provided in the transverse, longitudinal, and vertical directions and around the perimeter of the structure to tie everything together.

With the exception of prestressed concrete, approximate methods of frame analysis are permitted for typical construction.

LIVE LOAD

Some assumptions are allowed when considering live loads. Live loads are to be applied only to the floor or roof under consideration. The far ends of columns built integrally with the structure are considered fixed. Common practice is to assume that the arrangement of live loads is limited to combinations of factored dead loads on all spans with full factored live load on two adjacent span and factored dead load on all spans with full factored live load on alternate spans.

T-BEAMS

When working with T-beam construction, the flange and web are to be built integrally or otherwise effectively bonded together. The width of slab effective as a T-beam flange must not exceed one-quarter of the span length of the beam, and the effective overhanging flange width on each side of the web shall not exceed eight times the slab thickness and one-half the clear distance to the next web.

There are times when beams are placed with a slab on only one side. When this is the case, the effective overhanging flange width shall not exceed 1⁄12 the span length of the beam or six times the slab thickness or one-half the clear distance to the next web.

Isolated beams where T-shapes are used to provide a flange for additional compression area requires a flange thickness of not less than one-half the width of web and an effective flange width not more than four times the width of web.

JOIST CONSTRUCTION

What is joist construction? It is a monolithic combination of regularly spaced ribs and a top slab arranged to span in one direction or two orthogonal directions. Ribs are required to have a minimum width of four inches. The depth of the ribs cannot be more than 3.5 times the minimum width of the rib. Thirty inches of clear space is required between ribs.

If fillers are used, there must be a minimum of 1⁄12 the clear distance between ribs and not less than 1.5 inches. Slab thickness requires a minimum clear distance of one-twelfth the distance between the ribs and not less than 2 inches.

Concrete floor finishes can be considered part of the required cover or total thickness for nonstructural considerations.

FIRST STEP

What you have just read is a good first step toward understanding the general basis of the concrete code. There is a lot more in the code than what I have covered here, but this chapter has shown you a good cross section of what to expect when you dig into code requirements.

Math plays a substantial role in the concrete code. It involves many equations. Your codebook provides strong background for you to work with. Many of the symbols and abbreviations are defined in the front section of the code. Definitions of terms are clear. It will take time and some work to master the code, but you can pick up the codebook and begin to understand and use it immediately. I suggest that you do so.

During my 30 years in the construction business, I have seen numerous people fall far short of their potential in the trades. There have been countless reasons for this. Alcohol and drugs certainly are a factor in many failed careers. Old-fashioned laziness has cut many construction careers short. But the code and licensing requirements are responsible for holding a lot of people back.

Being fluent in the code makes you much more valuable on the job. Consider the plumbing or electrical trades. How much does an apprentice make per year? What is the expected income of a journeyman? Masters of the trade can expect the highest income and have the ability to open their own businesses. What separates the people? Experience is a factor. Being willing to learn their respective codes and pass the required licensing tests are what separates the highest earners from the lowest-paid employees.

Where do you want to be in the food chain? Are you happy raking gravel in slab beds? Does rolling reinforcement wire out and placing rebar excite you? All of this is necessary in a learning process, but most people would prefer to being wearing the white hat and calling the shots. If you want to be a supervisor or a business owner, concentrate on the code. Knowledge is what makes you more valuable.

Construction detail samples

The material on the following pages is courtesy of Faddis Concrete Products.

doi: 10.1016/B978-1-85617-549-4.00020-4

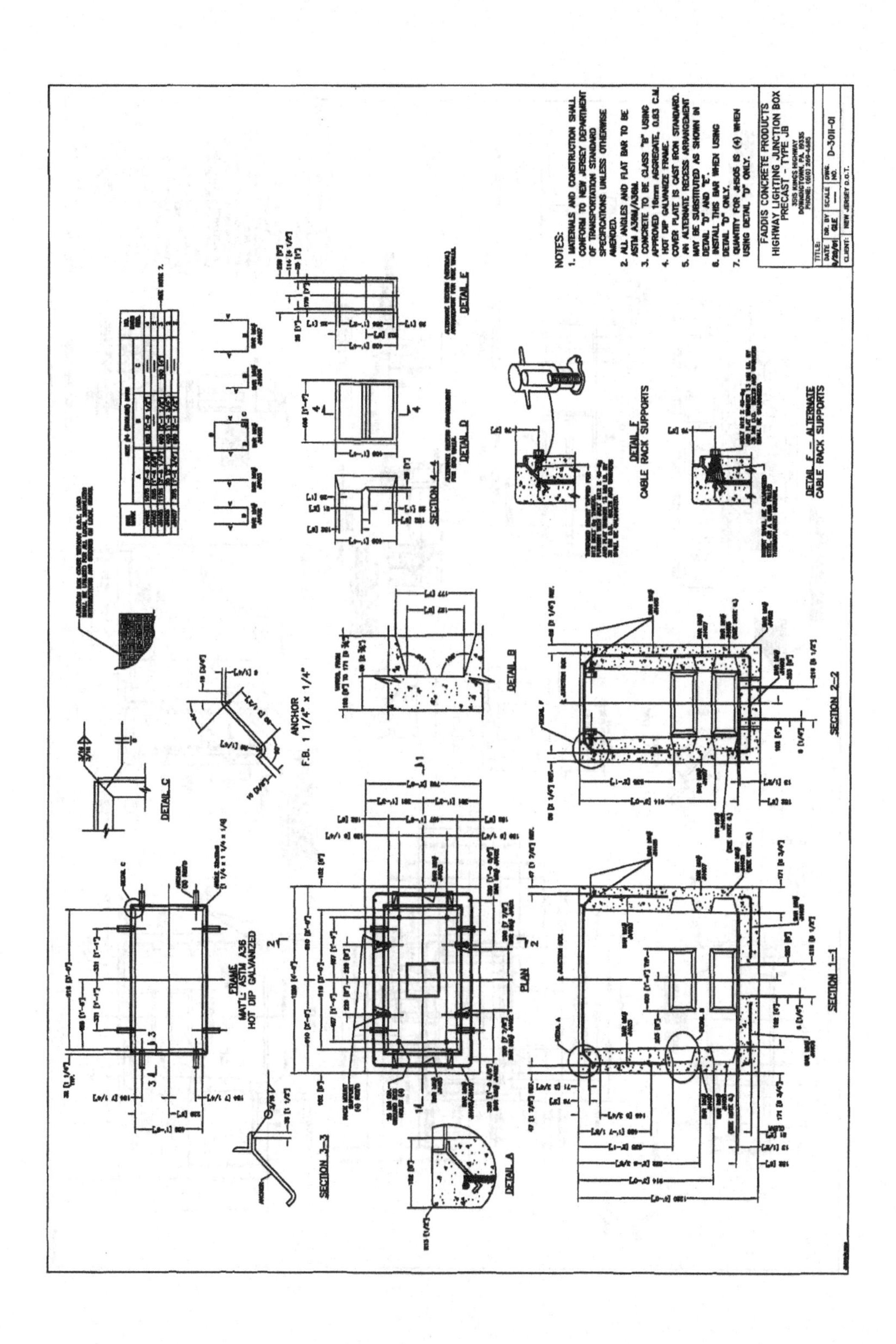
FRAME
MAT'L: ASTM A36
HOT DIP GALVANIZED
SECTION 3–3
ANCHOR
DETAIL C
ANCHOR
F.B. 1 1/4" x 1/4"
PLAN
DETAIL A
DETAIL B
SECTION 4–4
DETAIL D
DETAIL E
SECTION 1–1
SECTION 2–2
DETAIL F
CABLE RACK SUPPORTS
DETAIL F – ALTERNATE
CABLE RACK SUPPORTS
NOTES:
1. MATERIALS AND CONSTRUCTION SHALL CONFORM TO NEW JERSEY DEPARTMENT OF TRANSPORTATION STANDARD SPECIFICATIONS UNLESS OTHERWISE AMENDED.
2. ALL ANGLES AND FLAT BAR TO BE ASTM A36M/A36M.
3. CONCRETE TO BE CLASS "B" USING APPROVED 16mm AGGREGATE, 0.83 C.M.
4. HOT DIP GALVANIZE FRAME. COVER PLATE IS CAST IRON STANDARD.
5. AN ALTERNATE RECESS ARRANGEMENT MAY BE SUBSTITUTED AS SHOWN IN DETAIL "D" AND "E".
6. INSTALL THIS BAR WHEN USING DETAIL "D" ONLY.
7. QUANTITY FOR JH505 IS (4) WHEN USING DETAIL "D" ONLY.
FADDIS CONCRETE PRODUCTS
HIGHWAY LIGHTING JUNCTION BOX
PRECAST - TYPE JB
3515 KINGS HIGHWAY
DOWNINGTOWN, PA. 19335
PHONE: (610) 269-4685
TITLE:
DATE
DR. BY
GLE
SCALE
DWG. NO.
D-3011-01
CLIENT: NEW JERSEY D.O.T.

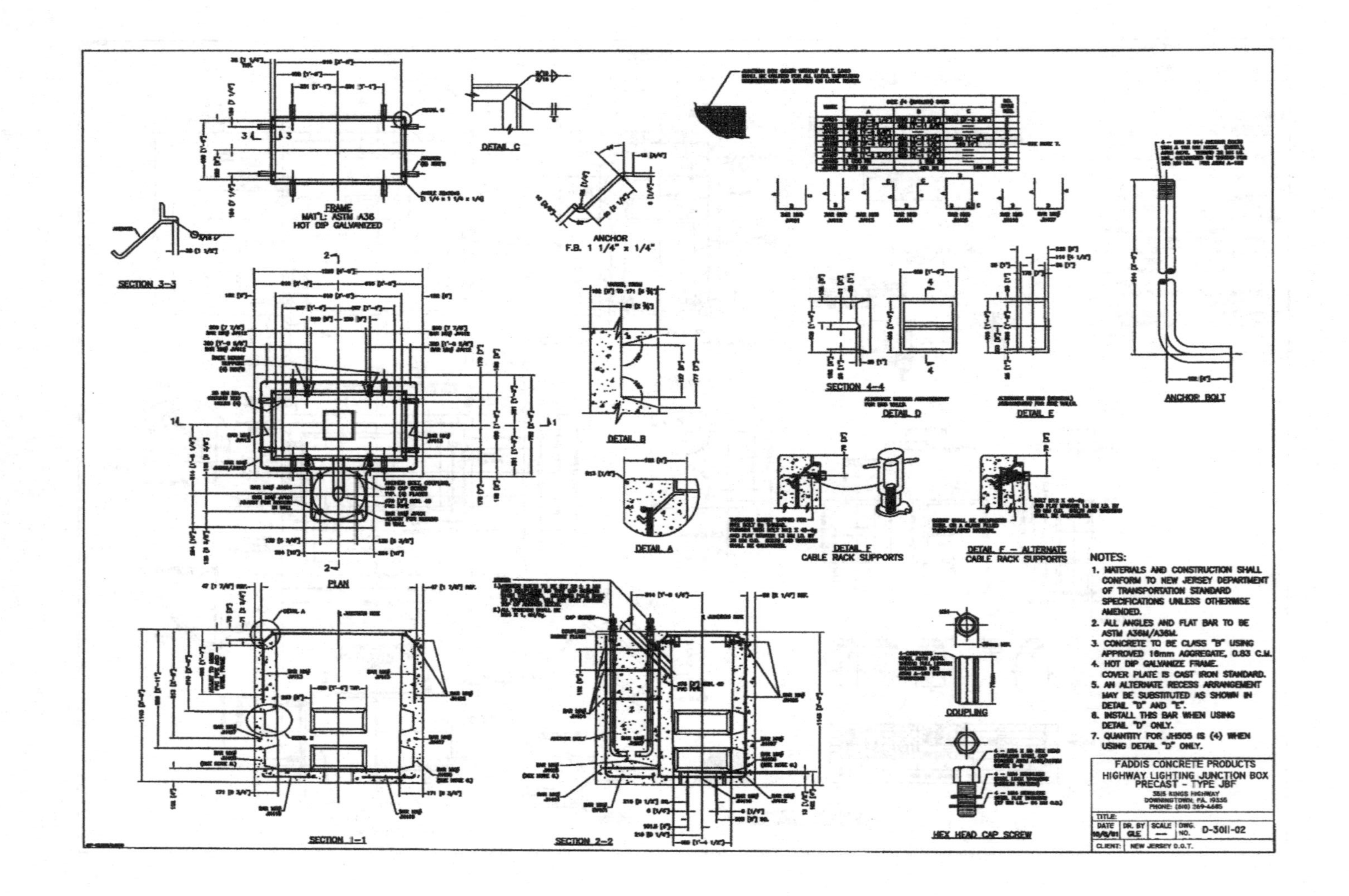
FRAME
MAT'L: ASTM A36
HOT DIP GALVANIZED
DETAIL C
ANCHOR
F.B. 1 1/4" x 1/4"
SECTION 3-3
PLAN
DETAIL B
DETAIL A
SECTION 4-4
DETAIL D
DETAIL E
ANCHOR BOLT
DETAIL F
CABLE RACK SUPPORTS
DETAIL F - ALTERNATE
CABLE RACK SUPPORTS
NOTES:
1. MATERIALS AND CONSTRUCTION SHALL CONFORM TO NEW JERSEY DEPARTMENT OF TRANSPORTATION STANDARD SPECIFICATIONS UNLESS OTHERWISE AMENDED.
2. ALL ANGLES AND FLAT BAR TO BE ASTM A36M/A36M.
3. CONCRETE TO BE CLASS "B" USING APPROVED 16mm AGGREGATE, 0.83 C.M.
4. HOT DIP GALVANIZE FRAME. COVER PLATE IS CAST IRON STANDARD.
5. AN ALTERNATE RECESS ARRANGEMENT MAY BE SUBSTITUTED AS SHOWN IN DETAIL "D" AND "E".
6. INSTALL THIS BAR WHEN USING DETAIL "D" ONLY.
7. QUANTITY FOR JH505 IS (4) WHEN USING DETAIL "D" ONLY.
COUPLING
HEX HEAD CAP SCREW
SECTION 1-1
SECTION 2-2
FADDIS CONCRETE PRODUCTS
HIGHWAY LIGHTING JUNCTION BOX
PRECAST - TYPE JBF
TITLE:
DATE
DR. BY
GLE
SCALE
DWG. NO.
D-3011-02
CLIENT:
NEW JERSEY D.O.T.

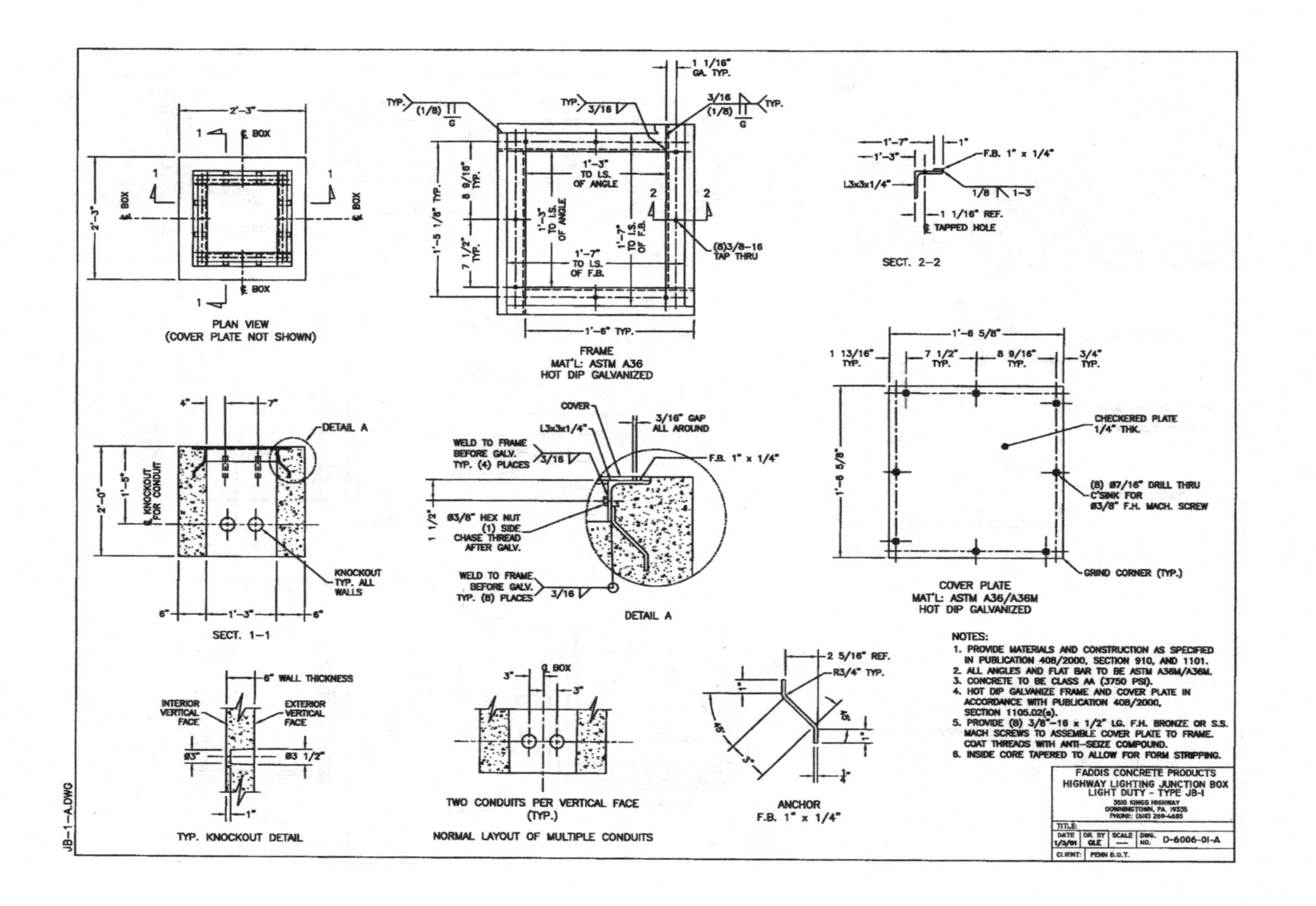
PLAN VIEW
(COVER PLATE NOT SHOWN)
FRAME
MAT'L: ASTM A36
HOT DIP GALVANIZED
SECT. 2-2
SECT. 1-1
DETAIL A
COVER PLATE
MAT'L: ASTM A36/A36M
HOT DIP GALVANIZED
CHECKERED PLATE
1/4" THK.
(8) Ø7/16" DRILL THRU
C'SINK FOR
Ø3/8" F.H. MACH. SCREW
GRIND CORNER (TYP.)
WELD TO FRAME
BEFORE GALV.
TYP. (4) PLACES
Ø3/8" HEX NUT
(1) SIDE
CHASE THREAD
AFTER GALV.
WELD TO FRAME
BEFORE GALV.
TYP. (8) PLACES
3/16" GAP
ALL AROUND
KNOCKOUT
TYP. ALL
WALLS
6" WALL THICKNESS
INTERIOR VERTICAL FACE
EXTERIOR VERTICAL FACE
TYP. KNOCKOUT DETAIL
TWO CONDUITS PER VERTICAL FACE
(TYP.)
NORMAL LAYOUT OF MULTIPLE CONDUITS
ANCHOR
F.B. 1" x 1/4"
NOTES:
1. PROVIDE MATERIALS AND CONSTRUCTION AS SPECIFIED IN PUBLICATION 408/2000, SECTION 910, AND 1101.
2. ALL ANGLES AND FLAT BAR TO BE ASTM A36M/A36M.
3. CONCRETE TO BE CLASS AA (3750 PSI).
4. HOT DIP GALVANIZE FRAME AND COVER PLATE IN ACCORDANCE WITH PUBLICATION 408/2000, SECTION 1105.02(e).
5. PROVIDE (8) 3/8"-16 x 1/2" LG. F.H. BRONZE OR S.S. MACH SCREWS TO ASSEMBLE COVER PLATE TO FRAME. COAT THREADS WITH ANTI-SEIZE COMPOUND.
6. INSIDE CORE TAPERED TO ALLOW FOR FORM STRIPPING.
FADDIS CONCRETE PRODUCTS
HIGHWAY LIGHTING JUNCTION BOX
LIGHT DUTY - TYPE JB-I
D-6006-01-A
CLIENT: PENN D.O.T.
JB-1-A.DWG

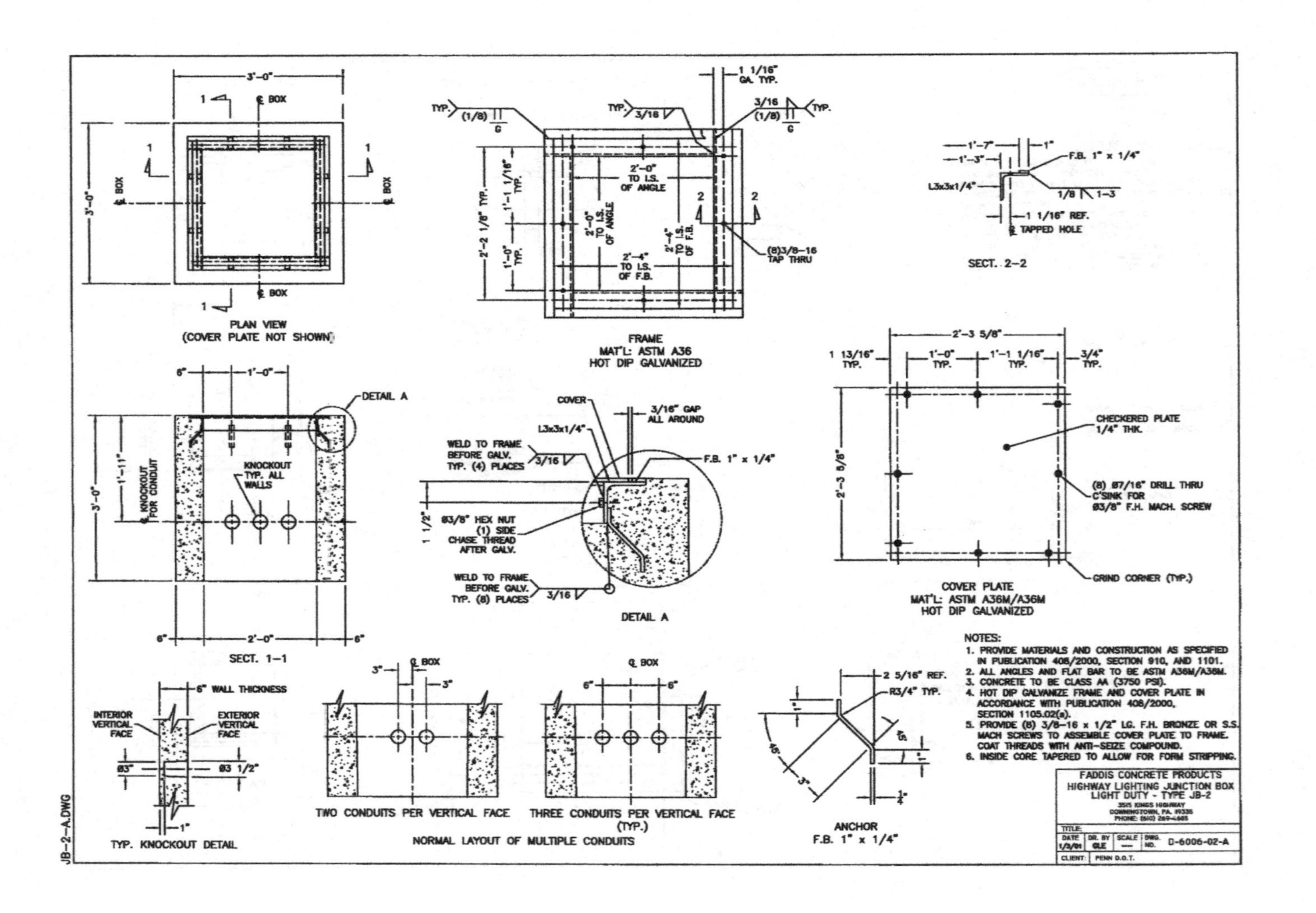
PLAN VIEW
(COVER PLATE NOT SHOWN)
FRAME
MAT'L: ASTM A36
HOT DIP GALVANIZED
2'-0" TO I.S. OF ANGLE
2'-4" TO I.S. OF F.B.
(8)3/8-16 TAP THRU
1 1/16" GA. TYP.
SECT. 2-2
F.B. 1" x 1/4"
L3x3x1/4"
1 1/16" REF.
℄ TAPPED HOLE
DETAIL A
KNOCKOUT TYP. ALL WALLS
℄ KNOCKOUT FOR CONDUIT
SECT. 1-1
COVER
3/16" GAP ALL AROUND
WELD TO FRAME BEFORE GALV. TYP. (4) PLACES
Ø3/8" HEX NUT (1) SIDE CHASE THREAD AFTER GALV.
WELD TO FRAME BEFORE GALV. TYP. (8) PLACES
COVER PLATE
MAT'L: ASTM A36M/A36M
HOT DIP GALVANIZED
CHECKERED PLATE 1/4" THK.
(8) Ø7/16" DRILL THRU C'SINK FOR Ø3/8" F.H. MACH. SCREW
GRIND CORNER (TYP.)
6" WALL THICKNESS
INTERIOR VERTICAL FACE
EXTERIOR VERTICAL FACE
TYP. KNOCKOUT DETAIL
TWO CONDUITS PER VERTICAL FACE
THREE CONDUITS PER VERTICAL FACE (TYP.)
NORMAL LAYOUT OF MULTIPLE CONDUITS
2 5/16" REF.
R3/4" TYP.
ANCHOR
F.B. 1" x 1/4"
NOTES:
1. PROVIDE MATERIALS AND CONSTRUCTION AS SPECIFIED IN PUBLICATION 408/2000, SECTION 910, AND 1101.
2. ALL ANGLES AND FLAT BAR TO BE ASTM A36M/A36M.
3. CONCRETE TO BE CLASS AA (3750 PSI).
4. HOT DIP GALVANIZE FRAME AND COVER PLATE IN ACCORDANCE WITH PUBLICATION 408/2000, SECTION 1105.02(s).
5. PROVIDE (8) 3/8-16 x 1/2" LG. F.H. BRONZE OR S.S. MACH SCREWS TO ASSEMBLE COVER PLATE TO FRAME. COAT THREADS WITH ANTI-SEIZE COMPOUND.
6. INSIDE CORE TAPERED TO ALLOW FOR FORM STRIPPING.
FADDIS CONCRETE PRODUCTS
HIGHWAY LIGHTING JUNCTION BOX
LIGHT DUTY - TYPE JB-2
D-6006-02-A
CLIENT: PENN D.O.T.
JB-2-A.DWG

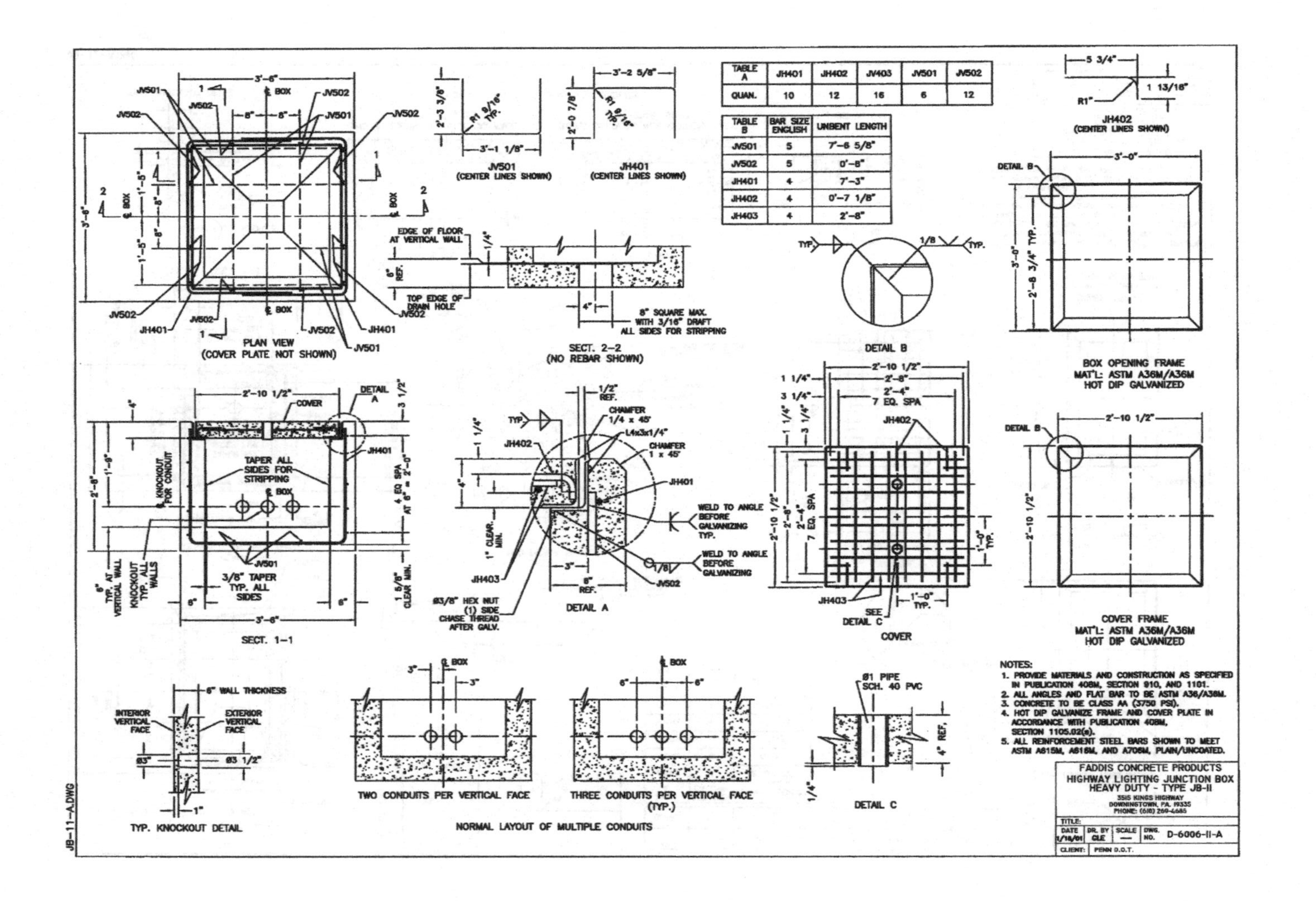

TABLE A	JH401	JH402	JV403	JV501	JV502
QUAN.	10	12	16	6	12

TABLE B	BAR SIZE ENGLISH	UNBENT LENGTH
JV501	5	7'-6 5/8"
JV502	5	0'-8"
JH401	4	7'-3"
JH402	4	0'-7 1/8"
JH403	4	2'-8"

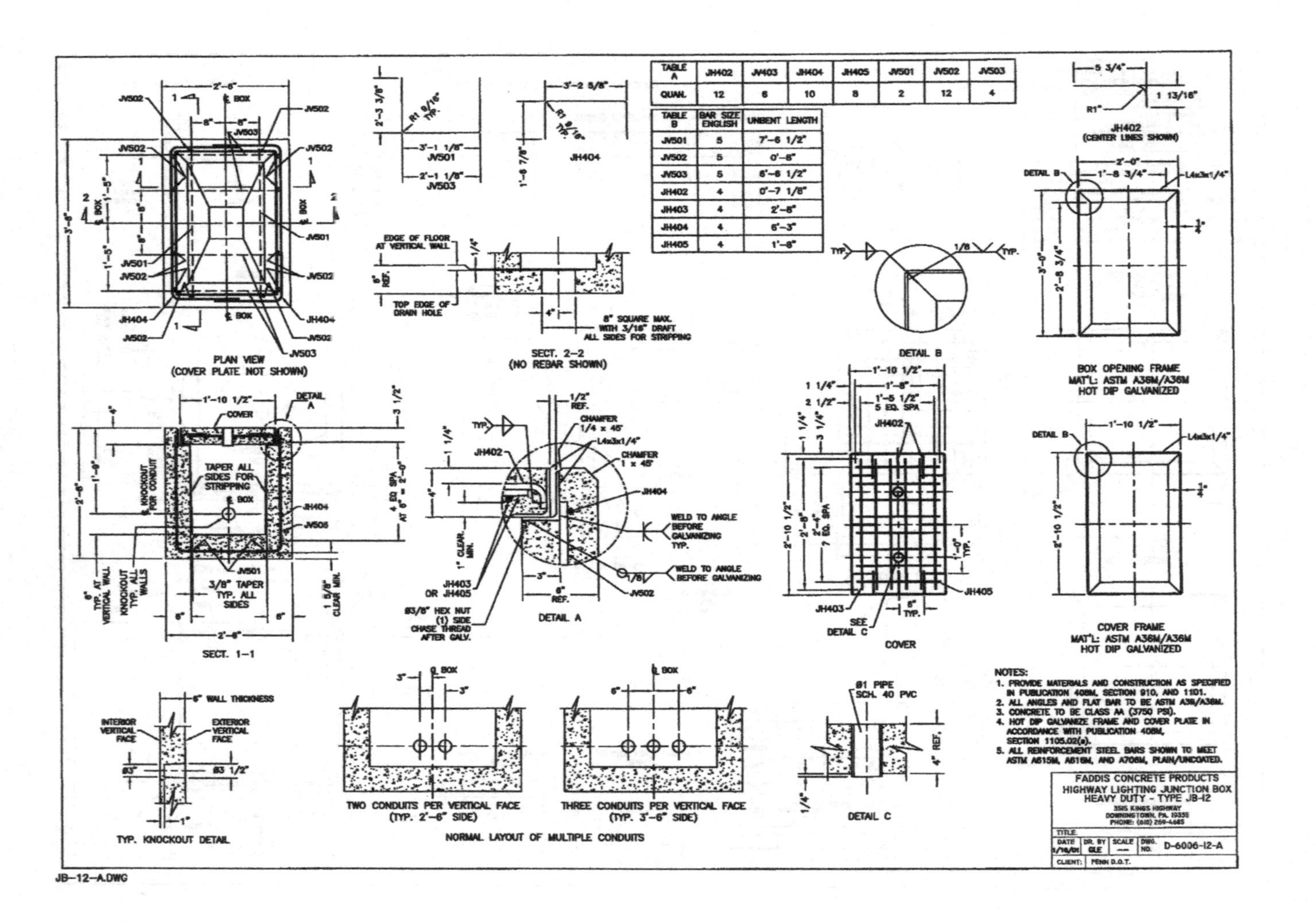
TABLE A | JH402 | JV403 | JH404 | JH405 | JV501 | JV502 | JV503
QUAN. | 12 | 6 | 10 | 8 | 2 | 12 | 4
TABLE B | BAR SIZE ENGLISH | UNBENT LENGTH
JV501 | 5 | 7'-6 1/2"
JV502 | 5 | 0'-8"
JV503 | 5 | 6'-6 1/2"
JH402 | 4 | 0'-7 1/8"
JH403 | 4 | 2'-8"
JH404 | 4 | 6'-3"
JH405 | 4 | 1'-8"
PLAN VIEW
(COVER PLATE NOT SHOWN)
SECT. 2-2
(NO REBAR SHOWN)
EDGE OF FLOOR AT VERTICAL WALL
TOP EDGE OF DRAIN HOLE
8" SQUARE MAX. WITH 3/16" DRAFT ALL SIDES FOR STRIPPING
DETAIL B
JH402
(CENTER LINES SHOWN)
BOX OPENING FRAME
MAT'L: ASTM A36M/A36M
HOT DIP GALVANIZED
SECT. 1-1
TAPER ALL SIDES FOR STRIPPING
3/8" TAPER TYP. ALL SIDES
KNOCKOUT FOR CONDUIT
KNOCKOUT TYP. ALL WALLS
6" TYP. AT VERTICAL WALL
1 5/8" CLEAR MIN.
DETAIL A
CHAMFER 1/4 x 45°
CHAMFER 1 x 45°
WELD TO ANGLE BEFORE GALVANIZING TYP.
WELD TO ANGLE BEFORE GALVANIZING
1" CLEAR. MIN.
JH403 OR JH405
Ø3/8" HEX NUT (1) SIDE CHASE THREAD AFTER GALV.
COVER
SEE DETAIL C
COVER FRAME
MAT'L: ASTM A36M/A36M
HOT DIP GALVANIZED
TYP. KNOCKOUT DETAIL
6" WALL THICKNESS
INTERIOR VERTICAL FACE
EXTERIOR VERTICAL FACE
TWO CONDUITS PER VERTICAL FACE
(TYP. 2'-6" SIDE)
THREE CONDUITS PER VERTICAL FACE
(TYP. 3'-6" SIDE)
NORMAL LAYOUT OF MULTIPLE CONDUITS
Ø1 PIPE SCH. 40 PVC
DETAIL C
NOTES:
1. PROVIDE MATERIALS AND CONSTRUCTION AS SPECIFIED IN PUBLICATION 408M, SECTION 910, AND 1101.
2. ALL ANGLES AND FLAT BAR TO BE ASTM A36/A36M.
3. CONCRETE TO BE CLASS AA (3750 PSI).
4. HOT DIP GALVANIZE FRAME AND COVER PLATE IN ACCORDANCE WITH PUBLICATION 408M, SECTION 1105.02(s).
5. ALL REINFORCEMENT STEEL BARS SHOWN TO MEET ASTM A615M, A616M, AND A706M, PLAIN/UNCOATED.
FADDIS CONCRETE PRODUCTS
HIGHWAY LIGHTING JUNCTION BOX
HEAVY DUTY - TYPE JB-12
D-6006-12-A
CLIENT: PENN D.O.T.
JB-12-A.DWG

GENERAL NOTES

1. PANEL SURFACES: ALUMINUM.

2. PANEL FILL: MINERAL WOOL

3. MATERIAL FINISH: POWDER COATING FOR ALUMINUM.

4. SOUND TRANSMISSION LOSS PER ASTM E90. SOUND ABSORPTION PER ASTM C423.

   SOUND TRANSMISSION COEFFICIENTS STC 30 ALUMINUM;

HZ	125	250	500	1000	2000	4000	
AF 1	16	16	27	39	40	35	FULL OCTAVE (ALUMINUM)

   SOUND ABSORBTION COEFFICIENTS:

HZ	125	250	500	1000	2000	4000	
AF 1	0.45	1.03	1.13	1.02	1.00	0.84	NRC 1.05

   POSTS, FOUNDATIONS AND ALL ASSOCIATED HARDWARE BY OTHERS.

5. DESIGN SPECIFICATIONS: 1989 AASHTO "GUIDE SPECIFICATIONS FOR THE STRUCTURAL DESIGN OF SOUND BARRIERS" AND DESIGN DIVISION 1 OF 1992 AASHTO
6. "SPECIFICATIONS FOR HIGHWAY BRIDGES", OR PER PROJECT REQUIREMENTS.

   STANDARD POST SPACING IS 8' 0" FOR BRIDGE MOUNT, 10'-0" FOR TYPICAL APPLICATIONS. POST SPACING UP TO 16'-0" CAN BE CONSIDERED FOR SPECIAL
7. APPLICATIONS.

8. FOR SOUND ABSORPTION INSTALLATION ON EXISTING REFLECTIVE PRECAST CONCRETE NOISE BARRIERS, OR OTHER HARD SURFACES, ATTACH PANELS AS SHOWN IN THE SOUND ABATEMENT RETROFIT DETAIL.

9. DIMENSIONS SHOWN IN ( ) ARE U.S. CUSTOMARY UNITS.

10. WEIGHT OF PANEL FOR DESIGN (INCLUDING ABSORPTIVE MATERIAL) IS:
    14.5 KG/SM (3.0 LBS/SF) – ALUMINUM
    PANEL WEIGHTS MAY VARY TO MEET ACOUSTICAL OR STRUCTURAL REQUIREMENTS.

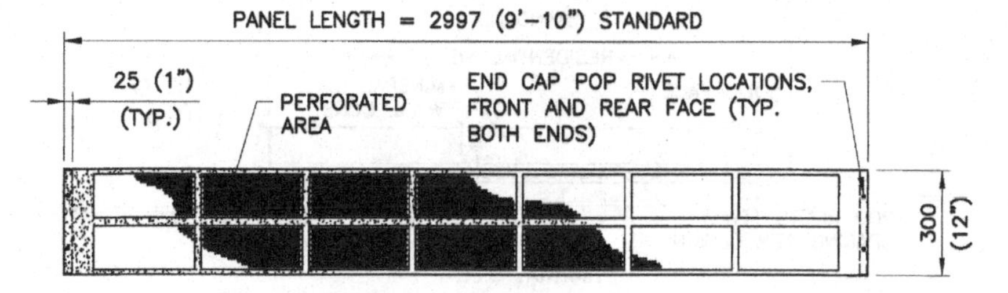

PANEL ELEVATION VIEW

ACOUSTAX	GENERALNOTES	PROJECT NO. PROJECT_NO	DRAWING NO.
	FADDIS CONCRETE PRODUCTS 3515 KINGS HIGHWAY DOWNINGTOWN, PA 19355 Voice: 610-269-4685 Fax: 610-873-8431	DATE: DATE DR.BY: DR_BY CH.BY: CH_BY SCALE: SCALE	1 OF 4

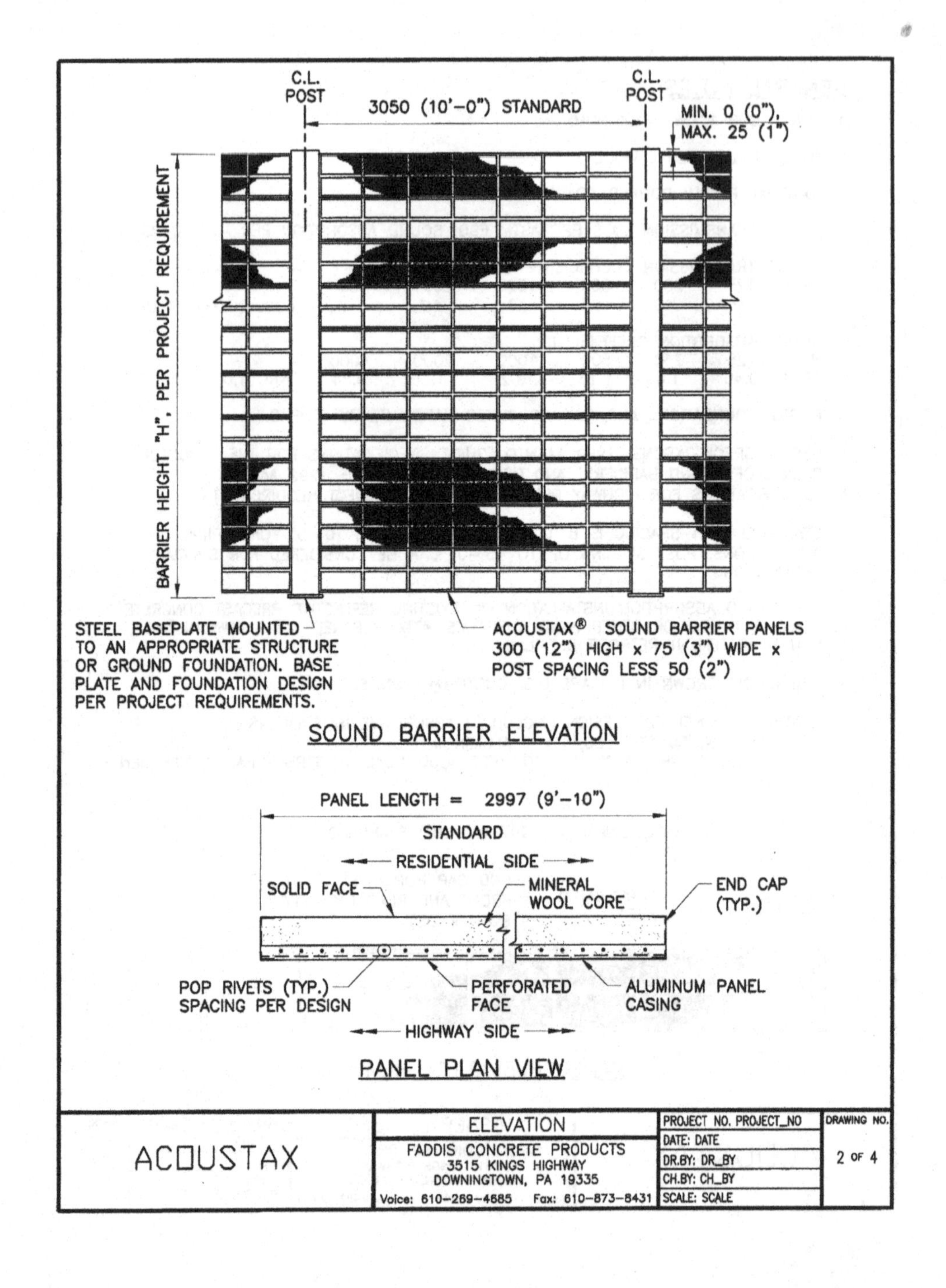
C.L. POST
3050 (10'-0") STANDARD
C.L. POST
MIN. 0 (0"), MAX. 25 (1")
BARRIER HEIGHT "H", PER PROJECT REQUIREMENT
STEEL BASEPLATE MOUNTED TO AN APPROPRIATE STRUCTURE OR GROUND FOUNDATION. BASE PLATE AND FOUNDATION DESIGN PER PROJECT REQUIREMENTS.
ACOUSTAX® SOUND BARRIER PANELS 300 (12") HIGH x 75 (3") WIDE x POST SPACING LESS 50 (2")
SOUND BARRIER ELEVATION
PANEL LENGTH = 2997 (9'-10")
STANDARD
RESIDENTIAL SIDE
SOLID FACE
MINERAL WOOL CORE
END CAP (TYP.)
POP RIVETS (TYP.) SPACING PER DESIGN
PERFORATED FACE
ALUMINUM PANEL CASING
HIGHWAY SIDE
PANEL PLAN VIEW
ACOUSTAX
ELEVATION
FADDIS CONCRETE PRODUCTS
3515 KINGS HIGHWAY
DOWNINGTOWN, PA 19335
Voice: 610-269-4685 Fax: 610-873-8431
PROJECT NO. PROJECT_NO
DATE: DATE
DR.BY: DR_BY
CH.BY: CH_BY
SCALE: SCALE
DRAWING NO.
2 OF 4

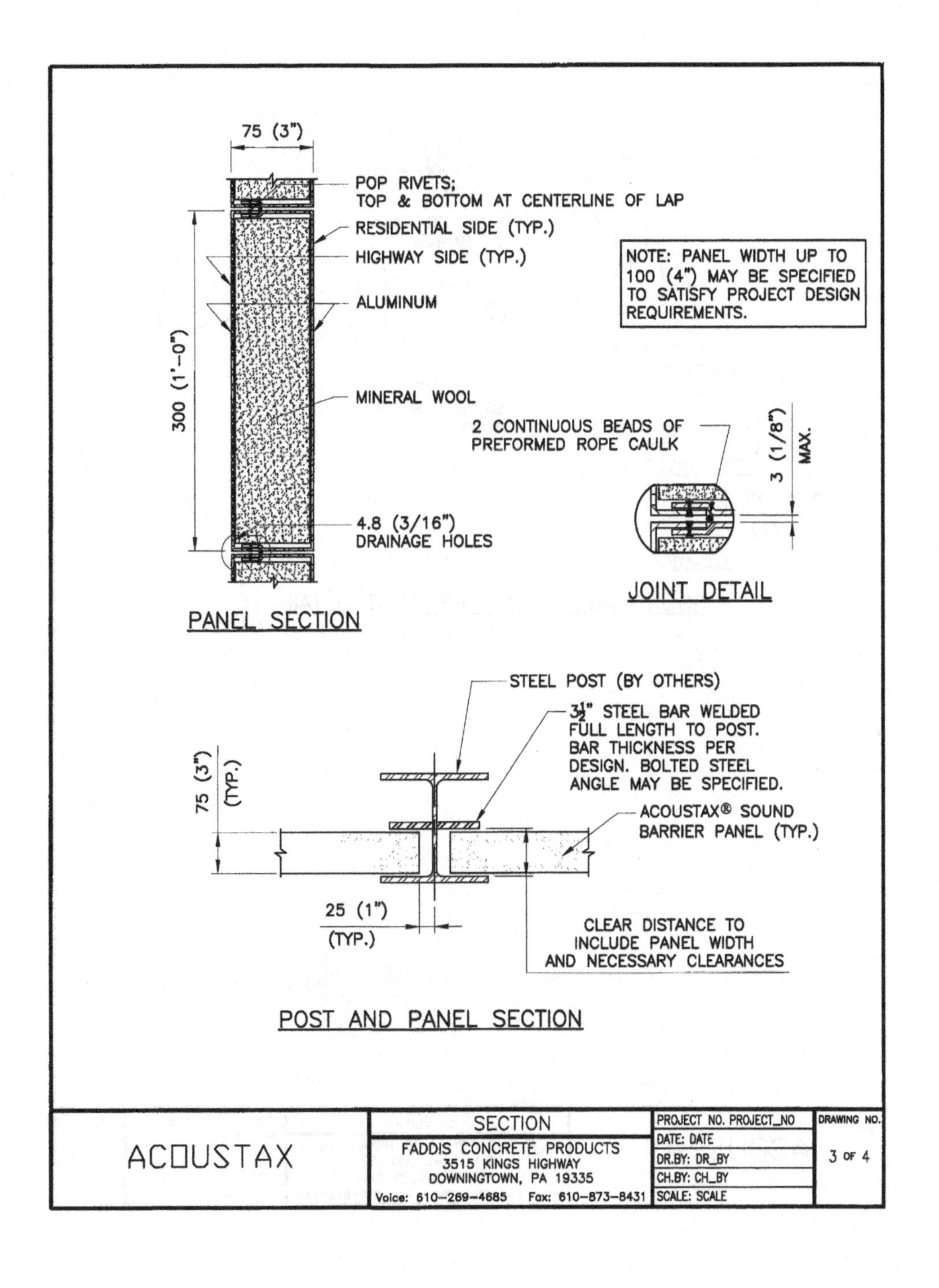

75 (3")
POP RIVETS;
TOP & BOTTOM AT CENTERLINE OF LAP
RESIDENTIAL SIDE (TYP.)
HIGHWAY SIDE (TYP.)
ALUMINUM
NOTE: PANEL WIDTH UP TO 100 (4") MAY BE SPECIFIED TO SATISFY PROJECT DESIGN REQUIREMENTS.
300 (1'-0")
MINERAL WOOL
2 CONTINUOUS BEADS OF PREFORMED ROPE CAULK
3 (1/8") MAX.
4.8 (3/16") DRAINAGE HOLES
JOINT DETAIL
PANEL SECTION
STEEL POST (BY OTHERS)
3½" STEEL BAR WELDED FULL LENGTH TO POST. BAR THICKNESS PER DESIGN. BOLTED STEEL ANGLE MAY BE SPECIFIED.
75 (3") (TYP.)
ACOUSTAX® SOUND BARRIER PANEL (TYP.)
25 (1") (TYP.)
CLEAR DISTANCE TO INCLUDE PANEL WIDTH AND NECESSARY CLEARANCES
POST AND PANEL SECTION
ACOUSTAX
SECTION
FADDIS CONCRETE PRODUCTS
3515 KINGS HIGHWAY
DOWNINGTOWN, PA 19335
Voice: 610-269-4685 Fax: 610-873-8431
PROJECT NO. PROJECT_NO
DATE: DATE
DR.BY: DR_BY
CH.BY: CH_BY
SCALE: SCALE
DRAWING NO.
3 OF 4

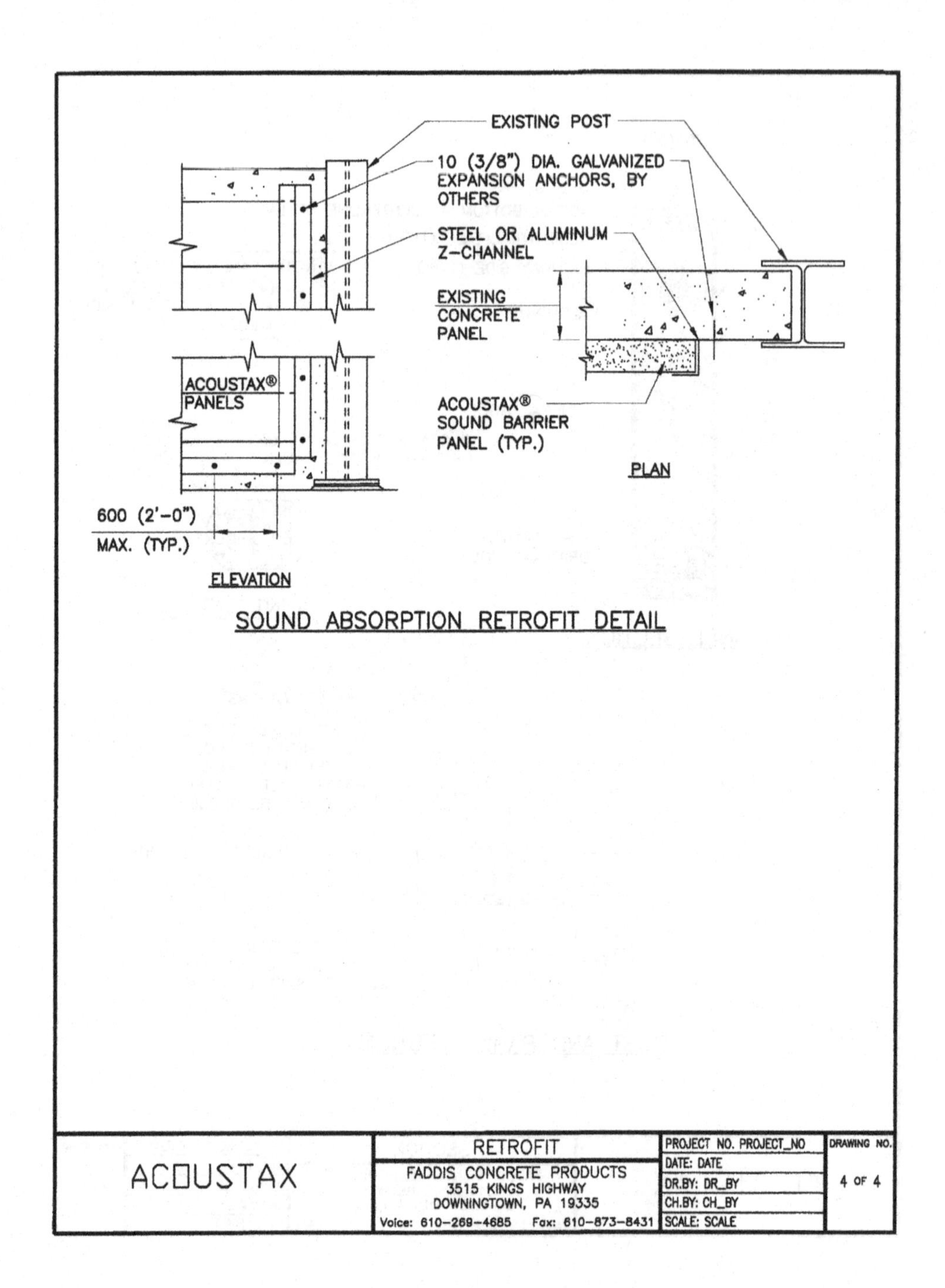

EXISTING POST
10 (3/8") DIA. GALVANIZED EXPANSION ANCHORS, BY OTHERS
STEEL OR ALUMINUM Z-CHANNEL
EXISTING CONCRETE PANEL
ACOUSTAX® PANELS
ACOUSTAX® SOUND BARRIER PANEL (TYP.)
PLAN
600 (2'-0") MAX. (TYP.)
ELEVATION
SOUND ABSORPTION RETROFIT DETAIL
ACOUSTAX
RETROFIT
FADDIS CONCRETE PRODUCTS
3515 KINGS HIGHWAY
DOWNINGTOWN, PA 19335
Voice: 610-269-4685 Fax: 610-873-8431
PROJECT NO. PROJECT_NO
DATE: DATE
DR.BY: DR_BY
CH.BY: CH_BY
SCALE: SCALE
DRAWING NO.
4 OF 4

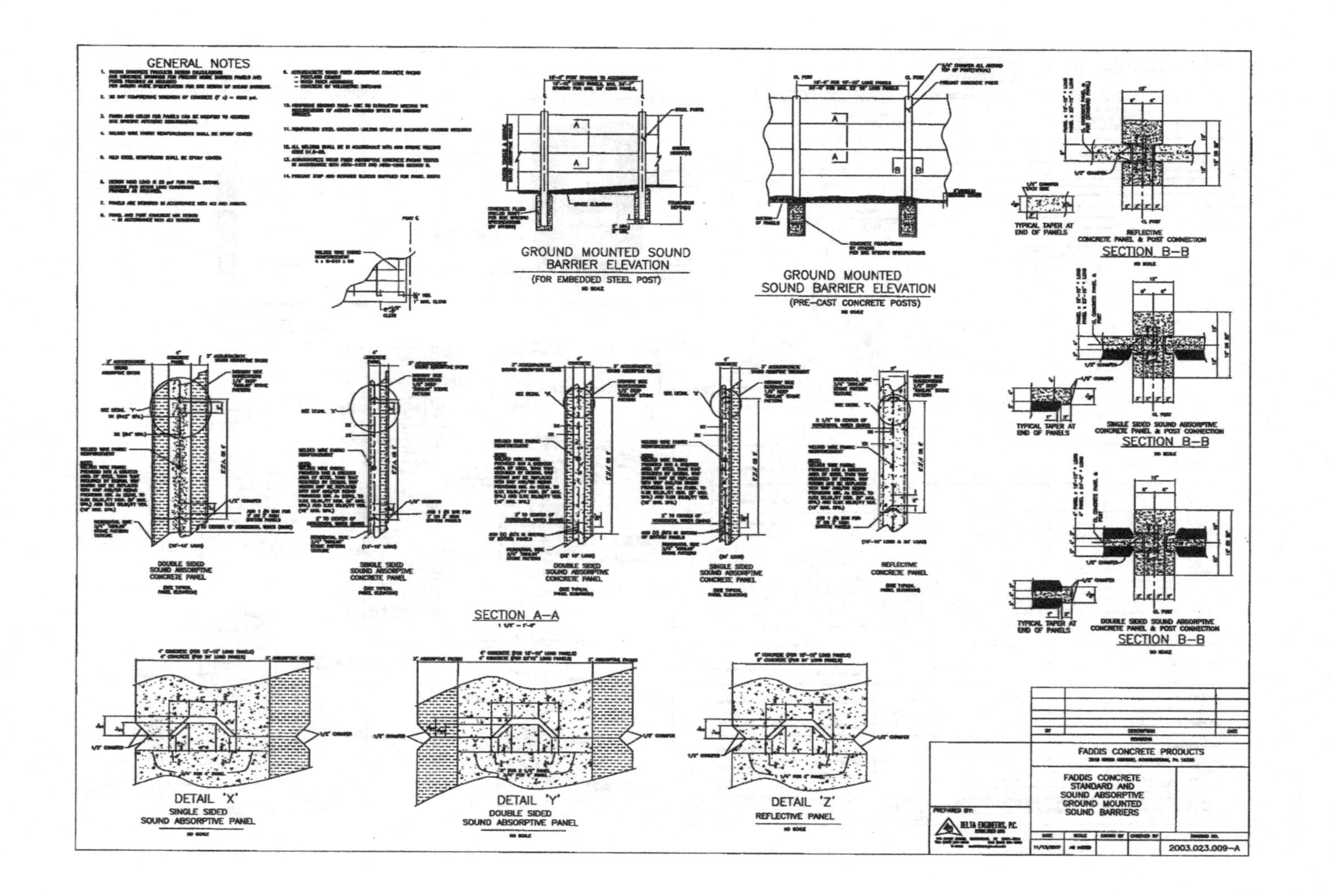
GENERAL NOTES
GROUND MOUNTED SOUND BARRIER ELEVATION
(FOR EMBEDDED STEEL POST)
GROUND MOUNTED SOUND BARRIER ELEVATION
(PRE-CAST CONCRETE POSTS)
TYPICAL TAPER AT END OF PANELS
REFLECTIVE CONCRETE PANEL & POST CONNECTION
SECTION B–B
TYPICAL TAPER AT END OF PANELS
SINGLE SIDED SOUND ABSORPTIVE CONCRETE PANEL & POST CONNECTION
SECTION B–B
TYPICAL TAPER AT END OF PANELS
DOUBLE SIDED SOUND ABSORPTIVE CONCRETE PANEL & POST CONNECTION
SECTION B–B
DOUBLE SIDED SOUND ABSORPTIVE CONCRETE PANEL
SINGLE SIDED SOUND ABSORPTIVE CONCRETE PANEL
DOUBLE SIDED SOUND ABSORPTIVE CONCRETE PANEL
SINGLE SIDED SOUND ABSORPTIVE CONCRETE PANEL
REFLECTIVE CONCRETE PANEL
SECTION A–A
DETAIL 'X'
SINGLE SIDED SOUND ABSORPTIVE PANEL
DETAIL 'Y'
DOUBLE SIDED SOUND ABSORPTIVE PANEL
DETAIL 'Z'
REFLECTIVE PANEL
PREPARED BY:
DELTA ENGINEERS, P.C.
FADDIS CONCRETE PRODUCTS
FADDIS CONCRETE STANDARD AND SOUND ABSORPTIVE GROUND MOUNTED SOUND BARRIERS
2003.023.009–A

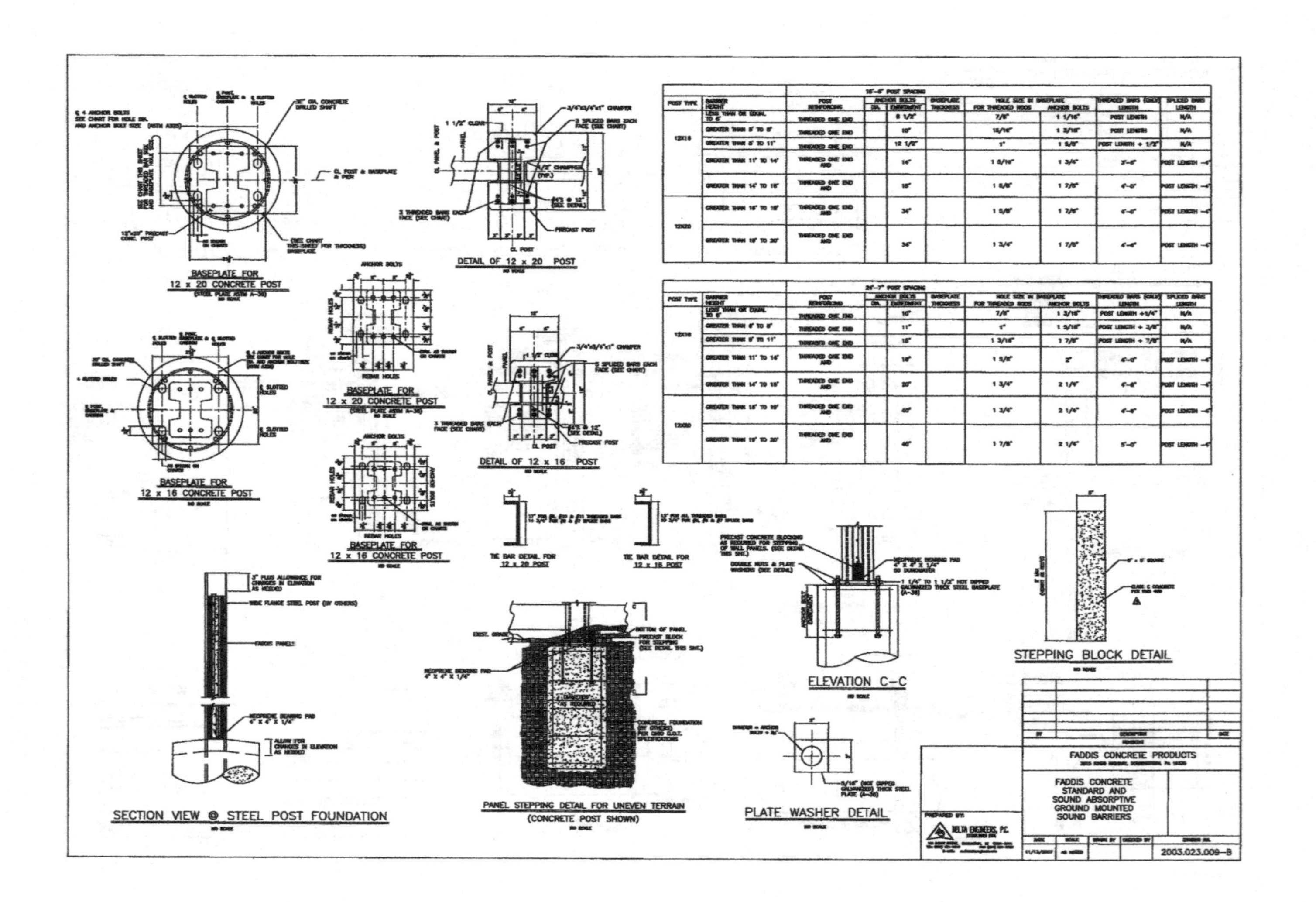

BASEPLATE FOR
12 x 20 CONCRETE POST
DETAIL OF 12 x 20 POST
BASEPLATE FOR
12 x 20 CONCRETE POST
DETAIL OF 12 x 16 POST
BASEPLATE FOR
12 x 16 CONCRETE POST
BASEPLATE FOR
12 x 16 CONCRETE POST
SECTION VIEW @ STEEL POST FOUNDATION
PANEL STEPPING DETAIL FOR UNEVEN TERRAIN
(CONCRETE POST SHOWN)
ELEVATION C–C
PLATE WASHER DETAIL
STEPPING BLOCK DETAIL
FADDIS CONCRETE PRODUCTS
FADDIS CONCRETE
STANDARD AND
SOUND ABSORPTIVE
GROUND MOUNTED
SOUND BARRIERS
2003.023.009–B

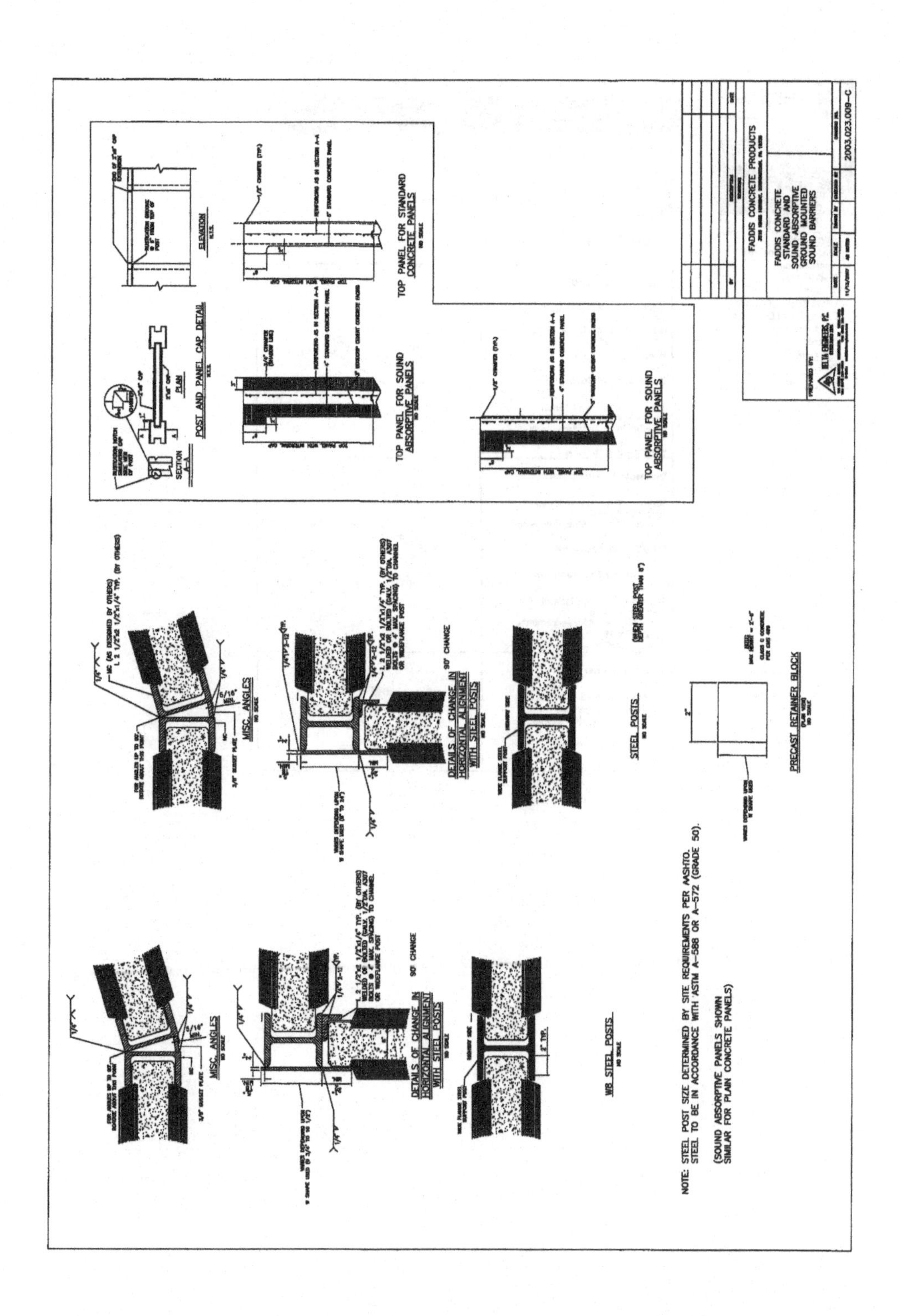
MISC. ANGLES
DETAILS OF CHANGE IN HORIZONTAL ALIGNMENT WITH STEEL POSTS
90° CHANGE
WB STEEL POSTS
STEEL POSTS
PRECAST RETAINER BLOCK
NOTE: STEEL POST SIZE DETERMINED BY SITE REQUIREMENTS PER AASHTO.
STEEL TO BE IN ACCORDANCE WITH ASTM A-588 OR A-572 (GRADE 50).
(SOUND ABSORPTIVE PANELS SHOWN
SIMILAR FOR PLAIN CONCRETE PANELS)
SECTION A-A
PLAN
POST AND PANEL CAP DETAIL
ELEVATION
TOP PANEL FOR SOUND ABSORPTIVE PANELS
TOP PANEL FOR STANDARD CONCRETE PANELS
TOP PANEL FOR SOUND ABSORPTIVE PANELS
FADDIS CONCRETE PRODUCTS
FADDIS CONCRETE
STANDARD AND
SOUND ABSORPTIVE
GROUND MOUNTED
SOUND BARRIERS
2003.023.009-C

DuraHold®

Single Depth

- System Components
- Specification
- Design
- Installation
- Cross Section Drawings
- Design Detail Drawings
- Case Studies
- Brochure
- Locate Manufacturer

Crib

Reinforced

DuraHold® System Components

Single Depth

DuraHold Coping Unit 305mm x 610mm x 1830mm (height x depth x length) Nominal Imperial Measurements: 12" x 24" x 72" Weight – 780 kg (1720 lbs) **DuraHold Half Coping Unit** 305mm x 610mm x 915mm (height x depth x length) Nominal Imperial Measurements: 12" x 24" x 36" Weight – 390 kg (860 lbs)	
DuraHold Standard Unit 305mm x 610mm x 1830mm (height x depth x length) Nominal Imperial Measurements: 12" x 24" x 72" Weight – 790 kg (1740 lbs) **DuraHold Half Standard Unit** 305mm x 610mm x 915mm (height x depth x length) Nominal Imperial Measurements: 12" x 24" x 36" Weight – 395 kg (870 lbs)	
DuraHold Corner Unit 305mm x 610mm x 1525mm (height x depth x length) Nominal Imperial Measurements: 12" x 24" x 60" Weight – 658 kg (1450 lbs)	

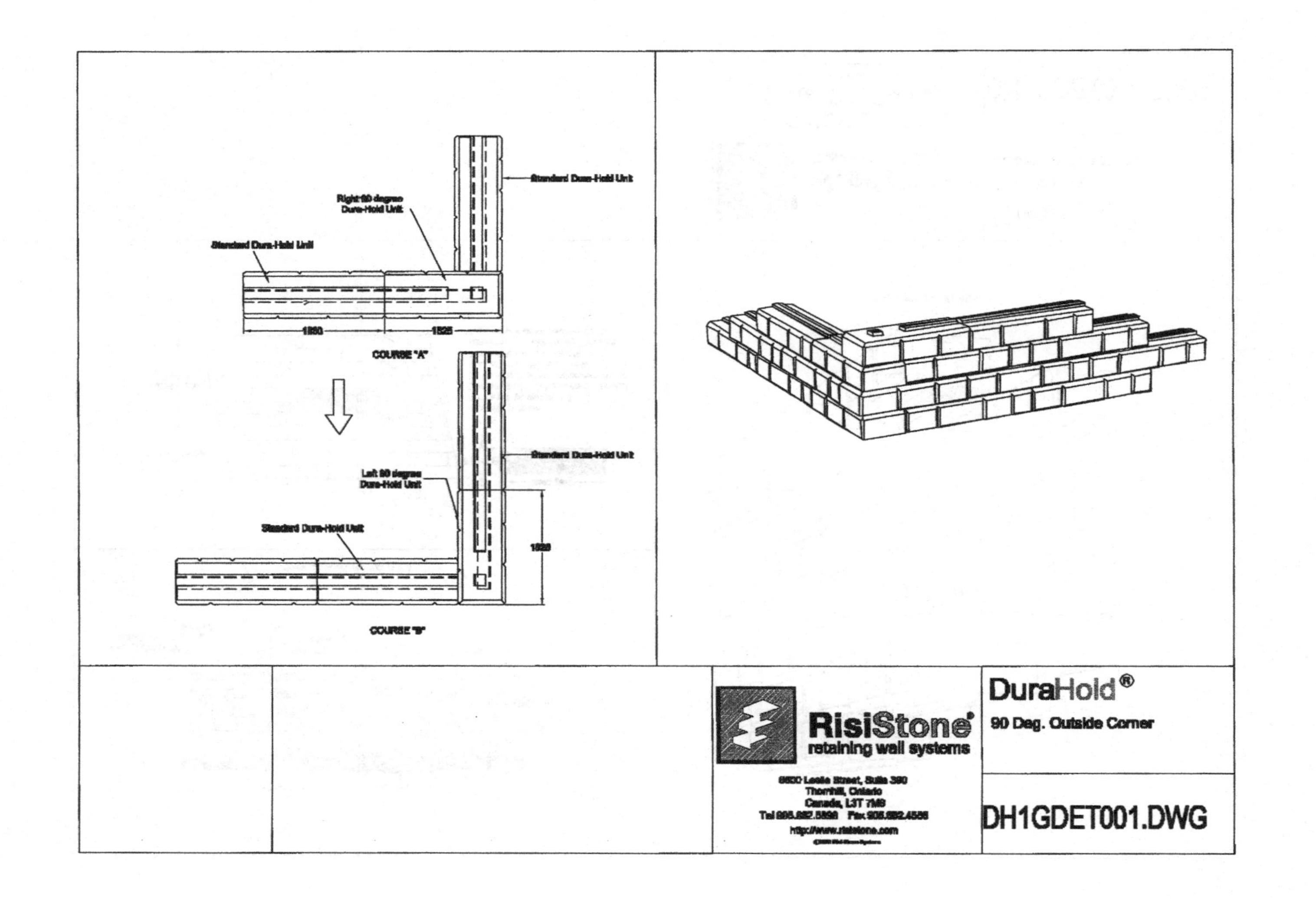
Standard Dura-Hold Unit
Right 90 degree Dura-Hold Unit
Standard Dura-Hold Unit
COURSE "A"
Standard Dura-Hold Unit
Left 90 degree Dura-Hold Unit
Standard Dura-Hold Unit
COURSE "B"
RisiStone®
retaining wall systems
DuraHold®
90 Deg. Outside Corner
DH1GDET001.DWG

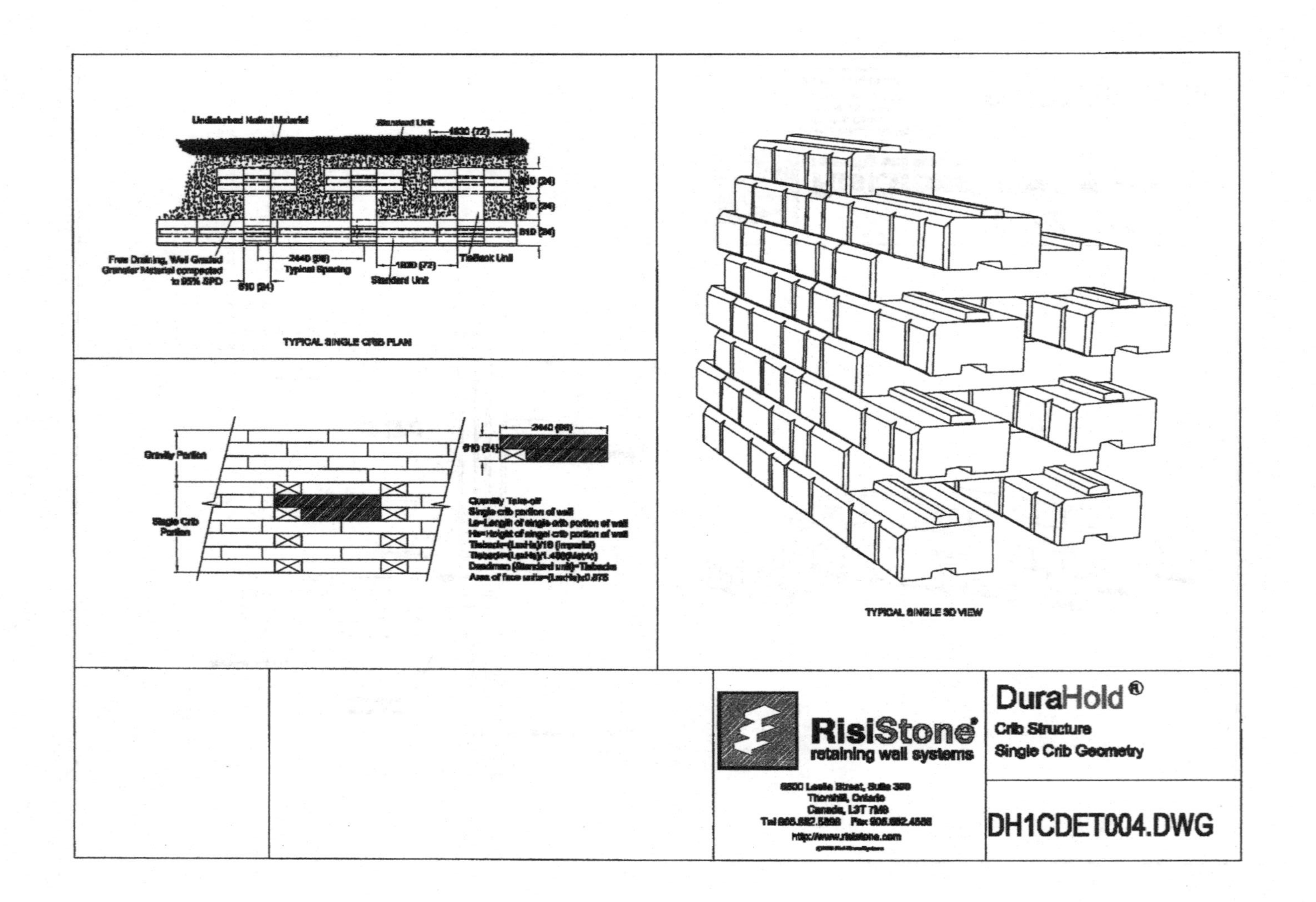
Undisturbed Native Material
Standard Unit
1830 (72)
610 (24)
610 (24)
610 (24)
Free Draining, Well Graded Granular Material compacted to 95% SPD
2440 (96)
Typical Spacing
1830 (72)
Standard Unit
Tieback Unit
610 (24)
TYPICAL SINGLE CRIB PLAN
Gravity Portion
Single Crib Portion
2440 (96)
610 (24)
Quantity Take-off
Single crib portion of wall
TYPICAL SINGLE 3D VIEW
RisiStone®
retaining wall systems
DuraHold®
Crib Structure
Single Crib Geometry
DH1CDET004.DWG

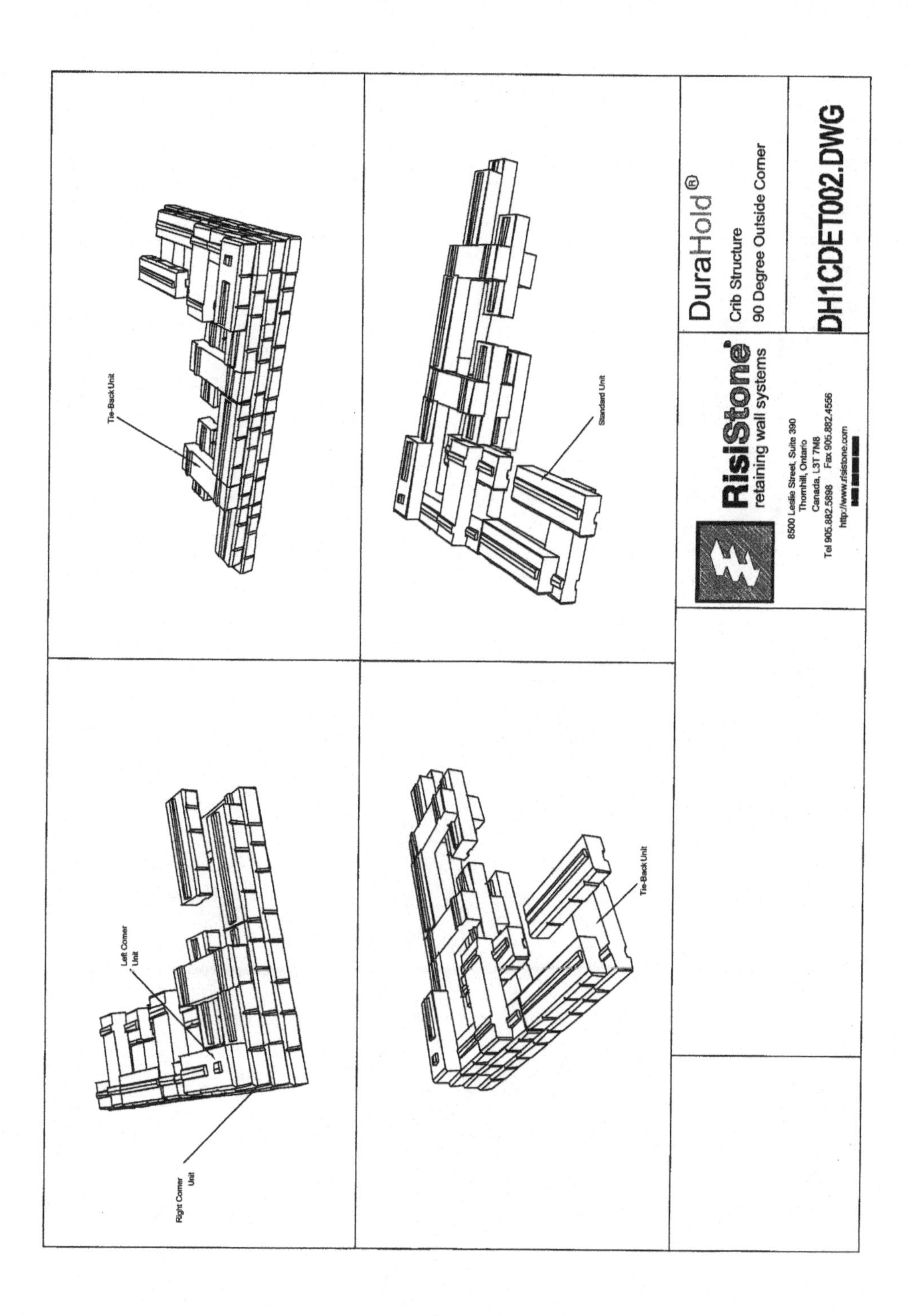
Tie-Back Unit
Standard Unit
Left Corner Unit
Right Corner Unit
Tie-Back Unit
DuraHold®
Crib Structure
90 Degree Outside Corner
DH1CDET002.DWG
RisiStone®
retaining wall systems
8500 Leslie Street, Suite 390
Thornhill, Ontario
Canada, L3T 7M8
Tel 905.882.5898 Fax 905.882.4556
http://www.risistone.com

APPENDIX

B Case study of highway pavement testing

The material on the following pages is courtesy of The United States Government.

doi: 10.1016/B978-1-85617-549-4.00021-6

SPECIAL REPORT 97-28

Current and Proposed Practices for Nondestructive Highway Pavement Testing

Maureen A. Kestler

November 1997

Abstract: In September 1994 the U.S. Army Cold Regions Research and Engineering Laboratory (CRREL) distributed a short survey on nondestructive testing practices to each of the 50 state Departments of Transportation (DOTs). The compilation of results constituted Phase I of a multiphase effort intended to lead toward the development of a method for optimizing falling weight deflectometer (FWD) test point spacing. Planned spatial statistical analyses on selected data sets will yield (site-specific) optimal FWD test point spacing for road network evaluation and pavement overlay design. Optimal FWD test point spacing reduces conservative overdesign due to undertesting and reduces overtesting. Both of these ultimately reduce expenditures. Although the above effort has not been completed, this interim report outlines the proposed process. Also included (and perhaps of more immediate interest to state DOTs) are direct survey facts and figures, including number of states with nondestructive testing (NDT) devices, average number of miles of annual overlay design, average number of miles of network/inventory testing, and back-calculation programs and overlay design procedures used. All facts and figures are generic and honor state anonymity.

How to get copies of CRREL technical publications:

Department of Defense personnel and contractors may order reports through the Defense Technical Information Center:

DTIC-BR SUITE 0944
8725 JOHN J KINGMAN RD
FT BELVOIRVA 22060-6218
Telephone 1 800 225 3842
E-mail help@dtic.mil
msorders@dtic.mil
WWW http://www.dtic.dla.mil/

All others may order reports through the National Technical Information Service:
NTIS
5285 PORT ROYAL RD
SPRINGFIELD VA 22161
Telephone 1 703 487 4650
1 703 487 4639 (TDD for the hearing-impaired)
E-mail orders@ntis.fedworld.gov
WWW http://www.fedworld.gov/ntis/ntishome.html

A complete list of all CRREL technical publications is available from
USACRREL (CECRL-LP)
72 LYME RD
HANOVER NH 03755-1290
Telephone 1 603 646 4338
E-mail techpubs@crrel.usace.army.mil

For information on all aspects of the Cold Regions Research and Engineering Laboratory, visit our World Wide Web site: http://www.crrel.usace.army.mil

Special Report 97-28

Current and Proposed Practices for Nondestructive Highway Pavement Testing

Maureen A. Kestler

November 1997

Prepared for
OFFICE OF THE CHIEF OF ENGINEERS

PREFACE

This report was prepared by Maureen A. Kestler, Research Civil Engineer, Civil Engineering Research Division, Research and Engineering Directorate, U.S. Army Cold Regions Research and Engineering Laboratory, Hanover, New Hampshire.

Funding for this work came from DA Project 4A762784AT42, Work Package 225, *Pavements in Cold Regions,* Work Unit CP-S01, *Mechanistic Pavement Design and Evaluation Methods for Cold Regions.*

The author thanks Keith Stebbings, Rosa Affleck, and Jack Bayer of CRREL for their assistance, and also thanks CRREL's editing and graphics departments.

Current and Proposed Practices for Nondestructive Highway Pavement Testing

MAUREEN A. KESTLER

OVERVIEW

In September 1994 the U.S.Army Cold Regions Research and Engineering Laboratory (CRREL) distributed a short survey on nondestructive testing practices to each of the 50 state Departments of Transportation (DOTs). The following report briefly summarizes state responses to questions regarding nondestructive testing (NDT) equipment used or owned, number of lane-miles tested annually, software and analytical tools utilized, and NDT test point spacing and configuration.

Compilation of survey results constituted Phase I of a multiphase effort leading toward development of a method for optimizing falling weight deflectometer (FWD) test point spacing. Long-range objectives are to assess national expenditures on NDT and to work in cooperation with selected state DOTs to determine whether present costs for overlay design and pavement evaluation could be reduced by the development of a computer program that continually assesses and updates in-situ variability, and recommends an optimal distance to the next FWD test point as data are collected in the field. This interim report[1] includes neither an analysis nor a final product, but rather summarizes survey results and oudines the theory and planned approach for computer program development.

[1]This report was written in response to numerous requests for a copy of the paper associated with a presentation titled "What Do DOTs Do with FWDs?", given at the FWD User's Group Meeting in Raleigh, North Carolina, in October 1995.

SURVEY RESULTS

NDT equipment

The NDT Practices Survey was distributed to the 50 state DOTs during the fall of 1994.Thirty-eight states replied, indicating a response rate of 76%. Of the 38 responding states, 21 states own (Dynatest) FWDs. Further investigation (Dynatest 1993) beyond survey results showed that, as of November 1993, six of the nonresponding states also owned Dynatest FWDs. Two states contract FWD work, six states own KUABs, one state owns a Mechanics Foundation JILS, three states own Road Raters, and four states continue to use Dynaflects. Several states own combinations of the above devices, e.g., one state owns two Dynatest FWDs and one KUAB, another state owns one Dynatest FWD and two Dynaflects, etc. Each of the NDT devices reported in this survey are briefly discussed in Appendix A (Smith and Lytton 1984).

NDT equipment uses and software/analytical tools used

Predominant uses for NDT equipment are pavement overlay design, pavement evaluation, network/inventory, research, void detection, and load transfer for portland cement concrete (PCC) pavements. Table 1 summarizes NDT software and analytical tools most commonly used by state DOTs. Figures la and lb graphically show the breakdown in methods/software for overlay design, evaluation, network, and project level usage. American Association of State Highway and Transportation Officials (AASHTO) guidelines and DARWIN,

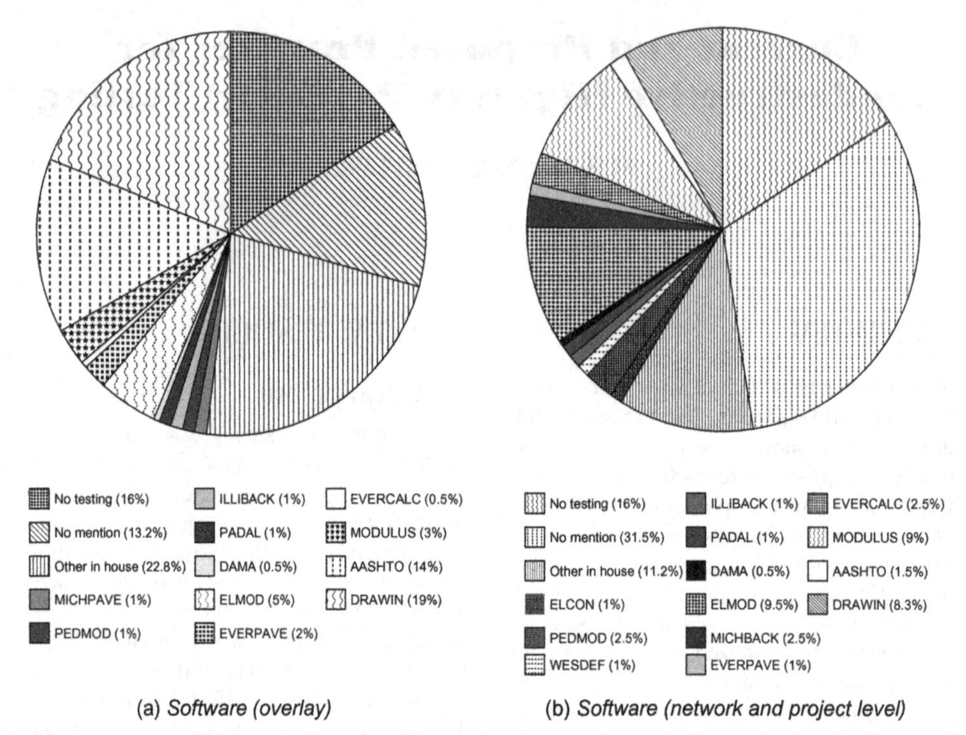

Figure 1. Breakdown of software used.

Table 1. Nondestructive testing software and analytical tools used by state DOTs.

	Evaluation/ network	*Overlay design*
DARWIN	X	X
AASHTO	X	X
MODULUS	X	X
EVERCALC	X	X
ELMOD	X	X
EVERPAVE	X	X
WESDEF	X	X
PADAL	X	X
ILLIBACK	X	X
ELCON	X	X
PEDMOD	X	X
DAMA	X	X
MICHBACK	X	
MICHPAVE		X
Other in-house programs	X	X

a computer-aided design method that uses AASHTO methods, provide the most frequently used overlay design technique of those reported. "In-house overlay design programs," ranging from sophisticated internally developed software to more simplistic spreadsheets, are also often used. (Note that weighted averages were assigned for figure development, i.e., states using only one software program were assigned a weight of one whereas states that specified three overlay design methods were assigned three weights of 1/3 each.)

Generally, FWDs are being used by state DOTs more for overlay design than for other purposes. However, DARWIN (and AASHTO), MODULUS, and ELMOD are the most commonly used methods for evaluation. Again, no individual in-house program is used across the board; nevertheless, a large percentage of states use their own software.

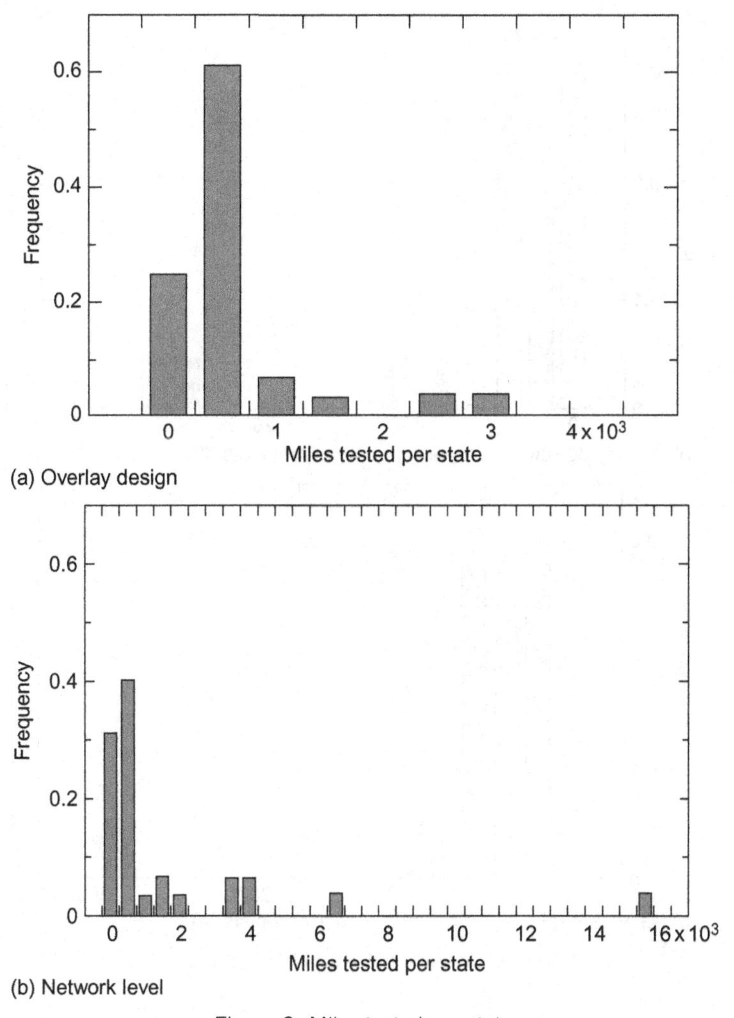

Figure 2. Miles tested per state.

Miles tested

Figures 2a and 2b show the distribution of lane-miles tested per state both for overlay design and at the network level, respectively. With a few exceptions, most states test fewer than 700 lane-miles per year for overlay design. Figure 2b shows that the vast majority of states test fewer than 1000 lane-miles per year at the network level. Note that these histograms indicate miles per state, not miles per piece of DOT equipment; thus, Figure 2b's outlier of approximately 15,000 miles per year corresponds to a state with an entire fleet of FWDs.

Spacing and configuration of NDT test points

Survey results indicated somewhat of a correlation between test pattern and "new and old"

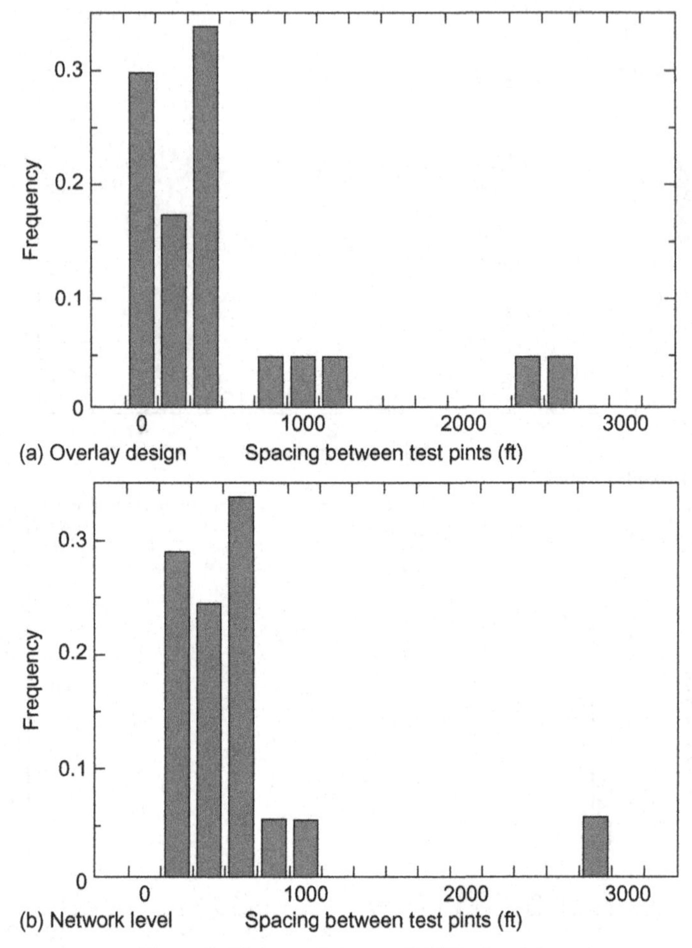

Figure 3. Distance between NDT test points.

FWD owners. Generally, new owners test in multiple locations (e.g., centerlines and wheelpaths) whereas veteran owners, for the most part, tend to test in just the right wheelpath. Spacing for purposes of network/inventory and overlay design tends to range from 100 to 1000 ft (Fig. 3a and 3b). Outlier reports of spacing for both purposes approach 3000 ft. Spacing could not always be quantified as a particular distance because several states indicated a minimum number of test points per project.

PLANS FOR OPTIMIZING TEST POINT SPACING

Pavement design and evaluation models and testing equipment continue to grow increasingly sophisticated, but only a limited amount of attention has been directed toward answering questions regarding the optimal number and location of FWD test points. Following completion of pavement strength variability analysis, we hope to minimize the cost for

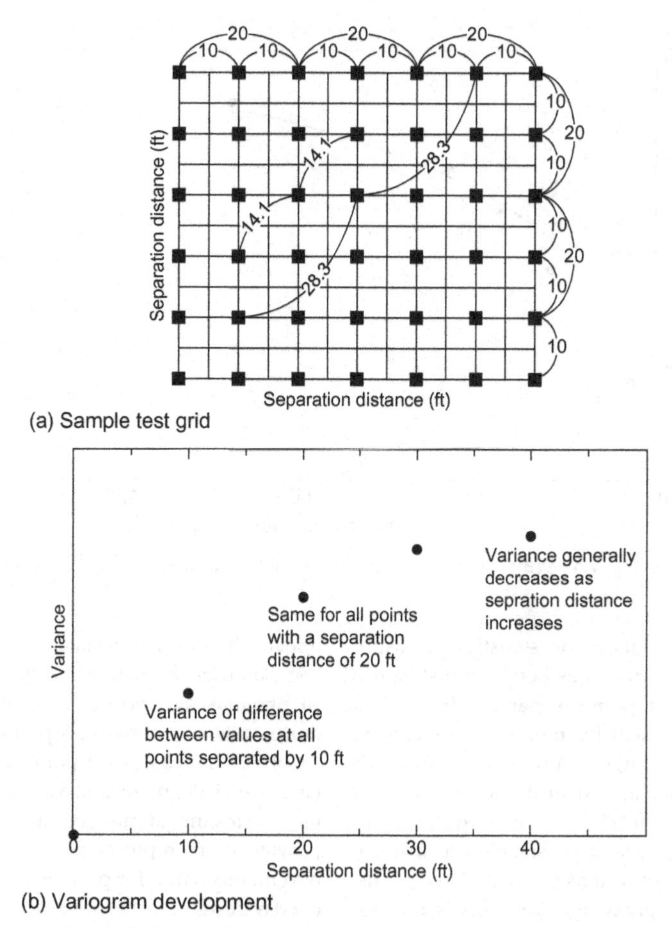

Figure 4. Test grid and associated variogram development.

overlay design and pavement evaluation by developing a computer program to optimize the number and location of FWD test points. Ideally, it would continually adjust the optimal distance to the next test point in real time as data are collected in the field. Based upon preliminary work, this continually adjusting optimal test point configuration computer program would maximize efficiency of FWD testing by 1) eliminating both undertesting and over-testing (thereby eliminating underdesign and over-design) , 2) minimizing lane closure time (thereby improving both employee and public safety), and 3) guaranteeing that adequate data are collected for overlay design and pavement evaluation.

Classical statistics can address random variability, but neglects relative positions of test points. However, there currently exist several less traditional mathematical models that can quantify spatial variability of pavement properties.

The following is a simplified look at a geo-statistical model that provides the basis for the proposed test point spacing optimization program.

Test points located close together (e.g., the 10-ft grid in Fig. 4a) yield similar test

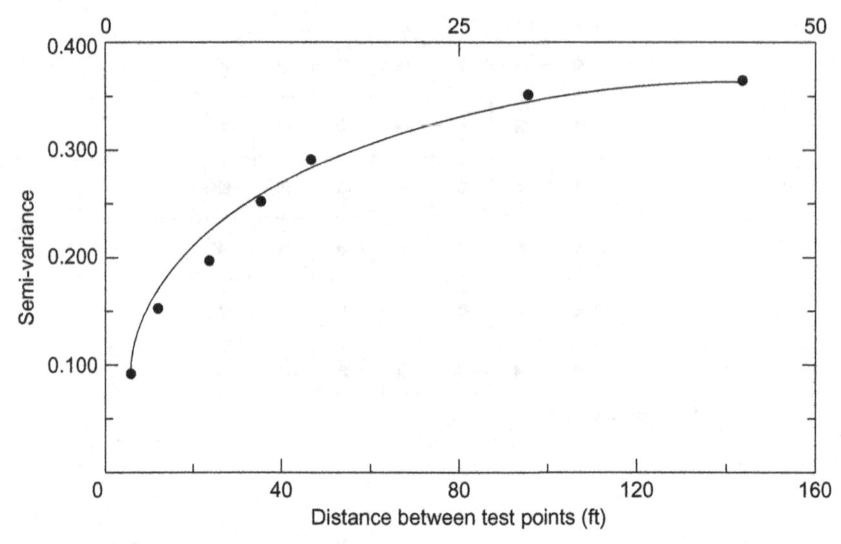

Figure 5. Variogram of normalized FWD center deflections, LV27 subgrade.

values. The variance (or statistical measure of spread) of differences between test values at pairs of test points separated by such a small distance will be minimal. This can be repeated for larger separation distances. For this particular test grid, the next closest spacing is 14.1 ft. The variance continues to increase up to a certain separation distance at which it levels out as shown in a geostatistical semivariogram (Fig. 4b). This is the distance beyond which the values (in this case, modulus or deflection) are no longer autocorrelated. Figure 5 shows the variogram corresponding to FWD data at a test cell at the Minnesota Road Research Program (Mn/ROAD) (Kestler et al. 1994). Points closer than approximately 150 ft are correlated; those spaced farther than 150 ft are independent of each other. While the variogram should define test point separation distances as outlined here, this analysis should remain invisible to the typical user.

There are currently many geostatistical software packages both available for purchase and in the public domain. They all analyze data at one point in time. We plan to modify an existing shareware package to continually update the optimal distance to the next FWD test point, as the data are collected, based on all previous data collected on that pavement during that test session. As pavement strength variability increases, test point separations distances will decrease and vice versa. This will reduce testing at unnecessary locations and provide more representative coverage of any pavement section for pavement evaluation and overlay design. Furthermore, this technique will probably minimize lane closure time, thereby improving both employee and public safety.

SUMMARY

Compilation of survey results from an NDT practices questionnaire distributed to state DOTs showed that Dynatest FWDs are by far the most popular nondestructive pavement testing device. The current uses for such FWD/NDT devices are pavement overlay design, pavement evaluation network/inventory research, void detection, and load transfer determination. Table 1 summarizes software and analytical tools used. AASHTO guidelines and DARWIN, which uses AASHTO methods, provide the most frequently used overlay design technique. In-house programs are also used quite often. DARWIN (and AASHTO), MODULUS, ELMOD, and in-house programs constitute the most commonly used methods for pavement evaluation.

Most states test fewer than 700 lane-miles per year for overlay design and fewer than 1000 lane-miles per year for network level testing. Test point spacing for both overlay design and network/ inventory ranges from 100 to 1000 ft, with outliers in both categories reaching 3000 ft.

As a follow-up to compiling and assessing survey results, we hope to minimize the cost for overlay design and pavement evaluation by developing a computer program that optimizes the number and location of FWD test points as data are collected in the field. The program will be based upon a mathematical model that enables quantification of spatial variability (of pavement stiffness). Based upon preliminary work, this continually adjusting optimal test point configuration program would optimize the FWD testing process by eliminating both undertesting and overtesting (thereby eliminating under- or overdesign), minimize lane closure time (thereby improving both employee and public safety), and guarantee that adequate data be collected for pavement evaluation and overlay design.

LITERATURE CITED

Dynatest (1993) *List of FWD Purchasers.* Dynatest, Stark, Florida.

Kestler, M.A., M.E. Harr, R.L. Berg, and D.M. Johnson (1994) Spatial variability of falling weight deflectometer data: A geostatistical analysis. In *Proceedings, Bearing Capacity of Roads and Airfields Conference, Minneapolis, Minnesota.*

Smith, R.E., and R.L. Lytton (1984) Synthesis study of NDT devices for use in overlay thickness design of flexible pavements. Report No. FHWA/ RD83-097, April.

APPENDIX A: NONDESTRUCTIVE PAVEMENT TESTING DEVICES

GENERAL INFORMATION

Historically, pavement deflection data have been collected by a variety of equipment that falls into four categories: 1) static beam deflection equipment, 2) automated beam deflection equipment, 3) steady-state dynamic deflection equipment, and 4) impulse deflection equipment (Smith and Lytton 1984).

The following sections provide a brief and simplistic description of the only two categories still used by responding state DOTs, impulse deflection equipment and steady-state dynamic deflection equipment. Although written several years ago, Report No. FHWA/RD83-097 (Smith and Lytton 1984) provides a comprehensive overview of all four categories of NDT equipment and is highly recommended for detailed descriptions.

IMPULSE DEFLECTION DEVICES

Impulse deflection equipment includes any testing device that applies an impulse load to the pavement surface. This is accomplished by lifting and dropping a mass from an adjustable height onto a buffer system that transmits the force through a loading plate to the pavement surface. The impulse load and resulting pavement response closely approximate a moving wheel load and associated pavement deflection. FWDs are impulse deflection devices.

Dynatest falling weight deflectometer

As was discussed in the main text, Dynatest FWDs are by far the NDT device most commonly used by state DOTs. Dynatest pavement testing equipment is mounted on a trailer, and can be towed by a van, truck, or automobile. Drop heights are varied to yield a desired load range. Pavement response is measured by (generally) seven velocity transducers located at desired distances from the center of the load plate. The complete operation, including raising and lowering the load plate and sensors, raising and dropping the mass, recording deflections, and signaling the operator when the system can be moved to the next site, is computer controlled.

KUAB falling weight deflectometer

Conceptually, the KUAB is very similar to the Dynatest FWD: pavement testing equipment is mounted in a towable trailer, an impulse load is applied to the pavement system through a buffer system and steel plate, and complete operation of the device and testing sequence is controlled by a computer housed in the towing vehicle. The primary difference is the addition of a segmented steel plate to more evenly distribute the load on uneven surfaces. Also, deflection is measured by seismic displacement transducers that are differential transformers, and the applied impulse load is longer than that of a Dynatest.

STEADY-STATE DYNAMIC DEFLECTION EQUIPMENT

Steady-state dynamic deflection equipment includes any pavement testing device that applies a dynamic force to produce a sinusoidal vibration in the pavement system. A static load is placed on the pavement surface, then a steady-state sinusoidal vibration is induced with a dynamic-force generator. The magnitude of peak-to-peak force is generally increased during testing. (The static load must also be adjusted accordingly to prevent the device from bouncing off the pavement surface.)

Dynaflect

Chronologically preceding FWDs, the Dynaflect is also trailer mounted and can be towed by any standard vehicle. The Dynaflect was one of the first types of steady-state dynamic deflection devices on the market.

A static load is applied to the pavement surface, then a sinusoidal vibration is applied through an unbalanced fly wheel system. Velocity transducers measure pavement deflection. Testing frequency and pavement response (deflections) are measured simultaneously. Before the introduction of FWDs, the Dynaflect was used for overlay design more than any other automated pavement testing device.

Road rater

Road Raters can be trailer mounted or incorporated within a vehicle such that towing is unnecessary. Load magnitudes vary for different models.

A load is applied to the pavement through a steel loading plate. The dynamic force is applied by a steel mass accelerated by a servo-controlled hydraulic actuator, and deflections are measured using four (or more) velocity transducers.

REPORT DOCUMENTATION PAGE

Form Approved
OMB No. 0704-0188

1. AGENCY USE ONLY (Leave blank)	2. REPORT DATE	3. REPORT TYPE AND DATES COVERED
	November 1997	

4. TITLE AND SUBTITLE

Current and Proposed Practices for Nondestructive Highway Pavement Testing

5. FUNDING NUMBERS

PR: 4A762784AT42
WP: 225
WU: CP-S01

6. AUTHORS

Maureen A. Kestler

7. PERFORMING ORGANIZATION NAME(S) AND ADDRESS(ES)

U.S. Army Cold Regions Research and Engineering Laboratory
72 Lyme Road
Hanover, New Hampshire 03755-1290

8. PERFORMING ORGANIZATION REPORT NUMBER

Special Report 97-28

9. SPONSORING/MONITORING AGENCY NAME(S) AND ADDRESS(ES)

U.S. Army Cold Regions Research and Engineering Laboratory
72 Lyme Road
Hanover, New Hampshire 03755-1290

10. SPONSORING/MONITORING AGENCY REPORT NUMBER

11. SUPPLEMENTARY NOTES

For conversion of SI units to non-SI units of measurement, consult ASTM Standard E380-93, *Standard Practice for Use of the International System of Units*, published by the American Society for Testing and Materials, 1916 Race St., Philadelphia, Pa. 19103.

12a. DISTRIBUTION/AVAILABILITY STATEMENT

Approved for public release; distribution is unlimited.

Available from NTIS, Springfield, Virginia 22161

12b. DISTRIBUTION CODE

13. ABSTRACT (*Maximum 200 words*)

In September 1994 the U.S. Army Cold Regions Research and Engineering Laboratory (CRREL) distributed a short survey on nondestructive testing practices to each of the 50 state Departments of Transportation (DOTs). The compilation of results constituted Phase I of a multiphase effort intended to lead toward the development of a method for optimizing falling weight deflectometer (FWD) test point spacing. Planned spatial statistical analyses on selected data sets will yield (site-specific) optimal FWD test point spacing for road network evaluation and pavement overlay design. Optimal FWD test point spacing reduces conservative overdesign due to undertesting and reduces overtesting. Both of these ultimately reduce expenditures. Although the above effort has not been completed, this interim report outlines the proposed process. Also included (and perhaps of more immediate interest to state DOTs) are direct survey facts and figures, including number of states with nondestructive testing (NDT) devices, average number of miles of annual overlay design, average number of miles of network/inventory testing, and back-calculation programs and overlay design procedures used. All facts and figures are generic and honor state anonymity.

14. SUBJECT TERMS

Back calculation, Falling weight deflectometer, FWD, NDT, Nondestructive testing, Pavement deflection analysis, Pavement evaluation, Pavement overlay design, Spatial variability

15. NUMBER OF PAGES

16. PRICE CODE

17. SECURITY CLASSIFICATION OF REPORT	18. SECURITY CLASSIFICATION OF THIS PAGE	19. SECURITY CLASSIFICATION OF ABSTRACT	20. LIMITATION OF ABSTRACT
UNCLASSIFIED	UNCLASSIFIED	UNCLASSIFIED	UL

NSN 7540-01-280-5500

Standard Form 298 (Rev. 2-89)
Prescribed by ANSI Std. Z39-18
298-102

APPENDIX

Helpful calculation methods

(Surface area ÷ R value) × (temperature inside – temperature outside)

Surface area of a material (in square feet) divided by its "R" value and multiplied by the difference in Fahrenheit degrees between inside and outside temperature equals heat loss in BTUs per hour.

FIGURE C.1 Calculating heat loss per hour with R-value.

- 3 feet of 1-in. pipe equal 1 square foot of radiation.
- 2⅓ linear feet of 1¼ in. pipe equal 1 square foot of radiation.
- Hot water radiation gives off 150 BTU per square foot of radiation per hour.
- Steam radiation gives off 240 BTU per square foot of radiation per hour.
- On greenhouse heating, figure ⅔ square foot of radiation per square foot of glass.
- One square foot of direct radiation condenses .25 pound of water per hour.

FIGURE C.2 Radiant heat facts.

$$L = \frac{144}{D \times 3.1414} \times R \div 12$$

D = O.D. of pipe

L = length of pipe needed in ft.

R = sq. ft. of radiation needed

FIGURE C.3 Formulas for pipe radiation.

doi: 10.1016/B978-1-85617-549-4.00024-1

The approximate weight of a piece of pipe can be determined by the following formulas:

Cast Iron Pipe: weight = $(A^2 - B^2) \times C \times .2042$

Steel Pipe: weight = $(A^2 - B^2) \times C \times .2199$

Copper Pipe: weight = $(A^2 - B^2) \times C \times .253$

A = outside diameter of the pipe in inches

B = inside diameter of the pipe in inches

C = length of the pipe in inches

FIGURE C.4 Finding the weight of piping.

The area of a pipe wall can be determined by the following formula:

Area of pipe wall = .7854 × [(O.D. × O.D.) – (I.D. × I.D.)]

FIGURE C.5 Finding the area of a pipe.

The formula for calculating expansion or contraction in plastic piping is: $L = Y \times \frac{T - F}{10} \times \frac{L}{100}$

L = Expansion in inches

Y = Constant factor expressing inches of expansion per 100°F temperature change per 100 ft. of pipe

T = Maximum temperature (0°F)

F = Minimum temperature (0°F)

L = Length of pipe run in feet

FIGURE C.6 Expansion in plastic piping.

The capacity of pipes is as the square of their diameters. Thus, doubling the diameter of a pipe increases its capacity four times.

FIGURE C.7 A piping fact.

Inch scale	Metric scale
1/16"	1:200

FIGURE C.8 Scale used for site plans.

Inch scale	Metric scale
¼"	1:50
⅛"	1:100

FIGURE C.9 Scales used for building plans.

To change	To	Multiply by
Inches	Feet	0.0833
Inches	Millimeters	25.4
Feet	Inches	12
Feet	Yards	0.3333
Yards	Feet	3
Square inches	Square feet	0.00694
Square feet	Square inches	144
Square feet	Square yards	0.11111
Square yards	Square feet	9
Cubic inches	Cubic feet	0.00058
Cubic feet	Cubic inches	1728
Cubic feet	Cubic yards	0.03703
Cubic yards	Cubic feet	27
Cubic inches	Gallons	0.00433
Cubic feet	Gallons	7.48
Gallons	Cubic inches	231
Gallons	Cubic feet	0.1337
Gallons	Pounds of water	8.33
Pounds of water	Gallons	0.12004
Ounces	Pounds	0.0625
Pounds	Ounces	16
Inches of water	Pounds per square inch	0.0361
Inches of water	Inches of mercury	0.0735
Inches of water	Ounces per square inch	0.578
Inches of water	Pounds per square foot	5.2
Inches of mercury	Inches of water	13.6
Inches of mercury	Feet of water	1.1333
Inches of mercury	Feet of water	0.4914
Ounces per square inch	Pounds per square inch	0.127
Ounces per square inch	Inches of mercury	1.733
Pounds per square inch	Inches of water	27.72
Pounds per square inch	Feet of water	2.31
Pounds per square inch	Inches of mercury	2.04
Pounds per square inch	Atmospheres	0.0681
Feet of water	Pounds per square inch	0.434
Feet of water	Pounds per square foot	62.5
Feet of water	Inches of mercury	0.8824
Atmospheres	Pounds per square inch	14.696
Atmospheres	Inches of mercury	29.92
Atmospheres	Feet of water	34
Long tons	Pounds	2240
Short tons	Pounds	2000
Short tons	Long tons	0.89295

FIGURE C.10 Useful multipliers.

To figure the final temperature when two different temperatures of water are mixed together, use the following formula:

$$\frac{(A \times C) + (B \times D)}{A + B}$$

A = Weight of lower temperature water

B = Weight of higher temperature water

C = Lower temperature

D = Higher temperature

FIGURE C.11 Temperature calculation.

Temperature can be expressed according to the Fahrenheit scale or the Celsius scale. To convert C to F or F to C, use the following formulas:

$°F = 1.8 \times °C + 32$

$°C = 0.55555555 \times °F - 32$

FIGURE C.12 Temperature conversion.

Deg. C. = Deg. F. – 32 ÷ 1.8

FIGURE C.13 Temperature conversion.

Deg. F. = Deg. C. × 1.8 + 32

FIGURE C.14 Temperature conversion.

Outside design temperature = Average of lowest recorded temperature in each month from October to March.

Inside design temperature = 70° Fahrenheit or as specified by owner.

A degree day is one day × the number of Fahrenheit degrees the mean temperature is below 65°F.
The number of degree days in a year is a good guideline for designing heating and insulation systems.

FIGURE C.15 Design temperature.

C	Base temperature	F
– 73	– 100	– 148
– 68	– 90	– 130
– 62	– 80	– 112
– 57	– 70	– 94
– 51	– 60	– 76
– 46	– 50	– 58
– 40	– 40	– 40
– 34.4	– 30	– 22
– 28.9	– 20	– 4
– 23.3	– 10	14
– 17.8	0	32
– 17.2	1	33.8
– 16.7	2	35.6
– 16.1	3	37.4
– 15.6	4	39.2
– 15.0	5	41.0
– 14.4	6	42.8
– 13.9	7	44.6
– 13.3	8	46.4
– 12.8	9	48.2
– 12.2	10	50
– 11.7	11	51.8
– 11.1	12	53.6
– 10.6	13	55.4
– 10.0	14	57.2

FIGURE C.16 Temperature conversion: – 100 to 30.

C	Base temperature	F
– 0.6	31	87.8
0	32	89.6
0.6	33	91.4
1.1	34	93.2
1.7	35	95
2.2	36	96.8
2.8	37	98.6
3.3	38	100.4
3.9	39	102.2
4.4	40	104.0
5	41	105.8
5.6	42	107.6
6.1	43	109.4
6.7	44	111.2
7.2	45	113.0
7.8	46	114.8
8.3	47	116.6
8.9	48	118.4
9.4	49	120.0
10	50	122.0
10.6	51	123.8
11.1	52	125.6
11.7	53	127.4
12.2	54	129.2
12.8	55	131.0

FIGURE C.17 Temperature conversion: 31 to 71.

C	Base temperature	F
22.2	72	161.6
22.8	73	163.4
23.3	74	165.2
23.9	75	167.0
24.4	76	168.8
25.0	77	170.6
25.6	78	172.4
26.1	79	174.2
26.7	80	176.0
27.8	81	177.8
28.3	82	179.6
28.9	83	181.4
29.4	84	183.2
30.0	85	185.0
30.6	86	186.8
31.1	87	188.6
31.7	88	190.4
32.2	89	192.2
32.8	90	194.0
33.3	91	195.8
33.9	92	197.6
34.4	93	199.4
35.0	94	201.2
35.6	95	203.0
36.1	96	204.8

FIGURE C.18 Temperature conversion: 72 to 212.

C	Base temperature	F
104	220	248
110	230	446
116	240	464
121	250	482
127	260	500
132	270	518
138	280	536
143	290	554
149	300	572
154	310	590
160	320	608
166	330	626
171	340	644
177	350	662
182	360	680
188	370	698
193	380	716
199	390	734
204	400	752
210	410	770
216	420	788
221	430	806
227	440	824
232	450	842
238	460	860

FIGURE C.19 Temperature conversion: 213 to 620.

C	Base temperature	F
332	630	1166
338	640	1184
343	650	1202
349	660	1220
354	670	1238
360	680	1256
366	690	1274
371	700	1292
377	710	1310
382	720	1328
388	730	1346
393	740	1364
399	750	1382
404	760	1400
410	770	1418
416	780	1436
421	790	1454
427	800	1472
432	810	1490
438	820	1508
443	830	1526
449	840	1544
454	850	1562
460	860	1580
466	870	1598

FIGURE C.20 Temperature conversion: 621 to 1000.

Quantity	Equals
60 seconds	1 minute
60 minutes	1 degree
360 degrees	1 circle

FIGURE C.21 Circular measure.

Function	Formula
Sine	$\sin = \frac{\text{side opposite}}{\text{hypotenuse}}$
Cosine	$\cos = \frac{\text{side opposite}}{\text{hypotenuse}}$
Tangent	$\tan = \frac{\text{side opposite}}{\text{hypotenuse}}$
Cosecant	$\csc = \frac{\text{hypotenuse}}{\text{side opposite}}$
Secant	$\sec = \frac{\text{hypotenuse}}{\text{side opposite}}$
Cotangent	$\cot = \frac{\text{hypotenuse}}{\text{side opposite}}$

FIGURE C.22 Trigonometry.

Multiply Length × Width × Thickness

Example: 50 ft. × 10 ft. × 8 in.

50' × 10' × $\frac{8}{12}$' = 333.33 cu. feet

To convert to cubic yards,
divide by 27 cu. ft. per cu.yd.

Example: 333.33 ÷ 27 = 12.35 cu. yd.

FIGURE C.23 Estimating volume.

Area of surface = Diameter × 3.1416 × length + area of the two bases

Area of base = Diameter × diameter × .7854

Area of base = Volume ÷ length

Length = Volume ÷ area of base

Volume = Length × area of base

Capacity in gallons = Volume in inches ÷ 231

Capacity of gallons = Diameter × diameter × length × .0034

Capacity in gallons = Volume in feet × 7.48

FIGURE C.24 Cylinder formulas.

Area = Short diameter × long diameter × .7854

FIGURE C.25 Ellipse calculation.

Area of surface = One half of circumference of base × slant height + area of base.

Volume = Diameter × diameter × .7854 × one-third of the altitude.

FIGURE C.26 Cone calculation.

Volume = Width × height × length

FIGURE C.27 Volume of a rectangular prism.

Area = Length × width

FIGURE C.28 Finding the area of a square.

Area = ½ perimeter of base × slant height + area of base

Volume = Area of base × ⅓ of the altitude

FIGURE C.29 Finding area and volume of a pyramid.

These comprise the numerous figures having more than four sides, names according to the number of sides, thus:

Figure	Sides
Pentagon	5
Hexagon	6
Heptagon	7
Octagon	8
Nonagon	9
Decagon	10

To find the area of a polygon: Multiply the sum of the sides (perimeter of the polygon) by the perpendicular dropped from its center to one of its sides, and half the product will be the area. This rule applies to all regular polygons.

FIGURE C.30 Polygons.

Area = Width of side × 2.598 × width of side

FIGURE C.31 Hexagons.

Area = Base × distance between the two parallel sides

FIGURE C.32 Parallelograms.

Area = Length × width

FIGURE C.33 Rectangles.

Area of surface = Diameter × diameter × 3.1416

Side of inscribed cube = Radius × 1.547

Volume = Diameter × diameter × diameter × .5236

FIGURE C.34 Spheres.

Area = One-half of height times base

FIGURE C.35 Triangles.

Area = One-half of the sum of the parallel sides × the height

FIGURE C.36 Trapezoids.

Volume = Width × height × length

FIGURE C.37 Cubes.

Circumference = Diameter × 3.1416

Circumference = Radius × 6.2832

Diameter = Radius × 2

Diameter = Square root of (area ÷ .7854)

Diameter = Square root of area × 1.1283

Diameter = Circumference × .31831

Radius = Diameter ÷ 2

Radius = Circumference × .15915

Radius = Square root of area × .56419

Area = Diameter × Diameter × .7854

Area = Half of the circumference × half of the diameter

Area = Square of the circumference × .0796

Arc length = Degrees × radius × .01745

Degrees of arc = Length ÷ (radius × .01745)

Radius of arc = Length ÷ (degrees ×.01745)

Side of equal square = Diameter × .8862

Side of inscribed square = Diameter × .7071

Area of sector = Area of circle × degrees of arc ÷ 360

FIGURE C.38 Formulas for a circle.

1. Circumference of a circle = π × diameter or 3.1416 × diameter
2. Diameter of a circle = Circumference × .31831
3. Area of a square = Length × width
4. Area of a rectangle = Length × width
5. Area of a parallelogram = Base × perpendicular height
6. Area of a triangle = ½ base × perpendicular height
7. Area of a circle = p × radius squared or diameter squared × .7854
8. Area of an ellipse = Length × width × .7854
9. Volume of a cube or rectangular prism = Length × width × height
10. Volume of a triangular prism = Area of triangle × length
11. Volume of a sphere = Diameter cubed × .5236 or (dia. × dia. × dia. × .5236)
12. Volume of a cone = π × radius square × ½ height
13. Volume of a cylinder = π × radius squared × height
14. Length of one side of a square × 1.128 = Diameter of an equal circle
15. Doubling the diameter of a pipe or cylinder increases its capacity 4 times
16. The pressure (in lbs. per sq. inch) of a column of water = Height of the column (in feet) × .434
17. The capacity of a pipe or tank (in U.S. gallons) =
    Diameter squared (in inches) × the length (in inches) × .0034
18. A gallon of water = 8⅓ lb. = 231 cu. inches
19. A cubic foot of water = 62½ lb. = 7½ gallons

FIGURE C.39 Useful formulas.

Number	Square
1	1
2	4
3	9
4	16
5	25
6	36
7	49
8	64
9	81
10	100
11	121
12	144
13	169
14	196
15	225
16	256
17	289
18	324
19	361
20	400
21	441
22	484
23	529
24	576
25	625
26	676
27	729
28	784
29	841
30	900
31	961
32	1024
33	1089
34	1156
35	1225
36	1296
37	1369
38	1444
39	1521
40	1600

Number	Square
41	1681
42	1764
43	1849
44	1936
45	2025
46	2116
47	2209
48	2304
49	2401
50	2500
51	2601
52	2704
53	2809
54	2916
55	3025
56	3136
57	3249
58	3364
59	3481
60	3600
61	3721
62	3844
63	3969
64	4096
65	4225
66	4356
67	4489
68	4624
69	4761
70	4900
71	5041
72	5184
73	5329
74	5476
75	5625
76	5776
77	5929
78	6084
79	6241
80	6400

Number	Square
81	6561
82	6724
83	6889
84	7056
85	7225
86	7396
87	7569
88	7744
89	7921
90	8100
91	8281
92	8464
93	8649
94	8836
95	9025
96	9216
97	9409
98	9604
99	9801
100	10000

FIGURE C.40 Squares of numbers.

Number	Square root
1	1.0000
2	1.4142
3	1.7321
4	2.0000
5	2.2361
6	2.4495
7	2.6458
8	2.8284
9	3.0000
10	3.1623
11	3.3166
12	3.4641
13	3.6056
14	3.7417
15	3.8730
16	4.0000
17	4.1231
18	4.2426
19	4.3589
20	4.4721
21	4.5826
22	4.6904
23	4.7958
24	4.8990
25	5.0000
26	5.0990
27	5.1962
28	5.2915
29	5.3852
30	5.4772
31	5.5678
32	5.6569
33	5.7446
34	5.8310
35	5.9161
36	6.0000
37	6.0828
38	6.1644
39	6.2450
40	6.3246
41	6.4031
42	6.4807
43	6.5574
44	6.6332
45	6.7082
46	6.7823
47	6.8557
48	6.9282
49	7.0000
50	7.0711
51	7.1414
52	7.2111
53	7.2801
54	7.3485
55	7.4162
56	7.4833
57	7.5498
58	7.6158
59	7.6811
60	7.7460
61	7.8102
62	7.8740
63	7.9373
64	8.0000
65	8.0623
66	8.1240
67	8.1854
68	8.2462
69	8.3066
70	8.3666
71	8.4261
72	8.4853
73	8.5440
74	8.6603
75	8.7178
76	8.7750
77	8.8318
78	8.8882
79	8.9443
80	9.0000
81	9.0554
82	9.1104
83	9.1104
84	9.1652
85	9.2195
86	9.2736
87	9.3274
88	9.3808
89	9.4340
90	9.4868
91	9.5394
92	9.5917
93	9.6437
94	9.6954
95	9.7468
96	9.7980
97	9.8489
98	9.8995
99	9.9499
100	10.0000

FIGURE C.41 Square roots of numbers.

Number	Cube
1	1
2	8
3	27
4	64
5	125
6	216
7	343
8	512
9	729
10	1000
11	1331
12	1728
13	2197
14	2477
15	3375
16	4096
17	4913
18	5832
19	6859
20	8000
21	9621
22	10648
23	12167
24	13824
25	15625
26	17576
27	19683
28	21952
29	24389
30	27000
31	29791
32	32768
33	35937
34	39304
35	42875
36	46656
37	50653
38	54872
39	59319
40	64000

Number	Cube
41	68921
42	74088
43	79507
44	85184
45	91125
46	97336
47	103823
48	110592
49	117649
50	125000
51	132651
52	140608
53	148877
54	157464
55	166375
56	175616
57	185193
58	195112
59	205379
60	216000
61	226981
62	238328
63	250047
64	262144
65	274625
66	287496
67	300763
68	314432
69	328500
70	343000
71	357911
72	373248
73	389017
74	405224
75	421875
76	438976
77	456533
78	474552
79	493039
80	512000

Number	Cube
81	531441
82	551368
83	571787
84	592704
85	614125
86	636056
87	658503
88	681472
89	704969
90	729000
91	753571
92	778688
93	804357
94	830584
95	857375
96	884736
97	912673
98	941192
99	970299
100	1000000

FIGURE C.42 Cubes of numbers.

Diameter	Area
⅛	0.0123
¼	0.0491
⅜	0.1104
½	0.1963
⅝	0.3068
¾	0.4418
⅞	0.6013
1	0.7854
1⅛	0.994
1¼	1.227
1⅜	1.484
1½	1.767
1⅝	2.073
1¾	2.405
1⅞	2.761
2	3.141
2¼	3.976
2½	4.908
2¾	5.939
3	7.068
3¼	8.295
3½	9.621
3¾	11.044
4	12.566
4½	15.904
5	19.635
5½	23.758
6	28.274
6½	33.183
7	38.484
7½	44.178
8	50.265
8½	56.745
9	63.617
9½	70.882
10½	86.59
11	95.03
11½	103.86
12	113.09
12½	122.71

Diameter	Area
13	132.73
13½	143.13
14	153.93
14½	165.13
15	176.71
15½	188.69
16	201.06
16½	213.82
17	226.98
17½	240.52
18	254.46
18½	268.8
19	283.52
19½	298.6
20	314.16
20½	330.06
21	346.36
21½	363.05
22	380.13
22½	397.6
23	415.47
23½	433.73
24	452.39
24½	471.43
25	490.87
26	530.93
27	572.55
28	615.75
29	660.52
30	706.89

FIGURE C.43 Area of circle.

Diameter	Circumference
⅛	.3927
¼	.7854
⅜	1.178
½	1.57
⅝	1.963
¾	2.356
⅞	3.748
1	3.141
1⅛	3.534
1¼	3.927
1⅜	4.319
1½	4.712
1⅝	5.105
1¾	5.497
1⅞	5.89
2	6.283
2¼	7.068
2½	7.854
2¾	8.639
3	9.424
3¼	10.21
3½	10.99
3¾	11.78
4	12.56
4½	14.13
5	15.7
5½	17.27
6	18.84
6½	20.42
7	21.99
7½	23.56
8	25.13
8½	26.7
9	28.27
9½	29.84
10	31.41
10½	32.98
11	34.55
11½	36.12
12	37.69

Diameter	Circumference
12½	39.27
13	40.84
13½	42.41
14	43.98
14½	45.55
15	47.12
15½	48.69
16	50.26
16½	51.83
17	53.4
17½	54.97
18	56.54
18½	58.11
19	59.69
19½	61.26
20	62.83
20½	64.4
21	65.97
21½	67.54
22	69.11
22½	70.68
23	72.25
23½	73.82
24	75.39
24½	76.96
25	78.54
26	81.68
27	84.82
28	87.96
29	91.1
30	94.24

FIGURE C.44 Circumference of a circle.

Decimal equivalent	Millimeters
.0625	1.59
.125	3.18
.1875	4.76
.25	6.35
.3125	7.94
.375	9.52
.4375	11.11
.5	12.70
.5625	14.29
.625	15.87
.6875	17.46
.75	19.05
.8125	20.64
.875	22.22
.9375	23.81
1.000	25.40

FIGURE C.45 Decimals to millimeters.

Inches	Decimal of a foot
⅛	.01042
¼	.02083
⅜	.03125
½	.04167
⅝	.05208
¾	.0625
⅞	.07291
1	08333
1⅛	.09375
1¼	.10417
1⅜	.11458
1½	.125
1⅝	.13542
1¾	.14583
1⅞	.15625
2	.16666
2⅛	.17708
2¼	.1875
2¾	.19792
2½	.20833
2⅝	.21875
2¾	.22917
2⅞	.23959
3	.2500

Note: To change inches to decimals of a foot, divide by 12.
To change decimals of a foot to inches, multiply by 12.

FIGURE C.46 Inches converted to decimals of feet.

Fractions	Decimal equivalent
1/16	.0625
⅛	.1250
3/16	.1875
¼	.2500
5/16	.3125
⅜	.3750
7/16	.4375
½	.5000
9/16	.5625
⅝	.6250
11/16	.6875
¾	.7500
13/16	.8125
⅞	.8750
15/16	.9375
1	1.000

FIGURE C.47 Fractions to decimals.

Fraction	Decimal
1/64	.015625
1/32	.03125
3/64	.046875
1/20	.05
1/16	.0625
1/13	.0769
5/64	.078125
1/12	.0833
1/11	.0909
3/32	.09375
1/10	.1
7/64	.109375
1/9	.111
1/8	.125
9/64	.140625
1/7	.1429
5/32	.15625
1/6	.1667
11/64	.171875
3/16	.1875
1/5	.2
13/64	.203125
7/32	.21875
15/64	.234375
1/4	.25
17/64	.265625
9/32	.28125
19/64	.296875

Fraction	Decimal
5/16	.3125
21/64	.328125
1/3	.333
11/32	.34375
23/64	.359375
3/8	.375
25/64	.390625
13/32	.40625
27/64	.421875
7/16	.4375
29/64	.453125
15/32	.46875
31/64	.484375
1/2	.5
33/64	.515625
17/32	.53125
35/64	.546875
9/16	.5625
37/64	.578125
19/32	.59375
39/64	.609375
5/8	.625
41/64	.640625
21/32	.65625
43/64	.671875
11/16	.6875
45/64	0.703125

FIGURE C.48 Decimals equivalents of fractions.

Minutes	Decimal of a degree
1	.0166
2	.0333
3	.05
4	.0666
5	.0833
6	.1
7	.1166
8	.1333
9	.15
10	.1666
11	.1833
12	.2
13	.2166
14	.2333
15	.25
16	.2666
17	.2833
18	.3
19	.3166
20	.3333
21	.35
22	.3666
23	.3833
24	.4000
25	.4166

FIGURE C.49 Minutes converted to decimal of a degree.

Fraction	Decimal
1/32	.03125
1/16	.0625
3/32	.09375
1/8	.125
5/32	.15625
3/16	.1875
7/32	.21875
1/4	.25
9/32	.28125
5/16	.3125
11/32	.34375
3/8	.375
13/32	.40625
7/16	.4375
15/32	.46875
1/2	.5
17/32	.53125
9/16	.5625
19/32	.59375
5/8	.625
21/32	.65625
11/16	.6875
23/32	.71875
3/4	.75
25/32	.78125
13/16	.8125
27/32	.84375
7/8	.875
29/32	.90625
15/16	.9375
31/32	.96875
1	1.000

FIGURE C.50 Decimal equivalents of an inch.

Inches	Decimal of an inch
1/64	0.015625
1/32	0.03125
3/64	0.046875
1/32	0.03125
3/64	0.046875
1/16	0.0625
5/64	0.078125
3/32	0.09375
7/64	0.109375
1/8	0.125
9/64	0.140625
5/32	0.15625
11/64	0.171875
3/16	0.1875
12/64	0.203125
7/32	0.21875
15/64	0.234375
1/4	0.25
17/64	0.265625
9/32	0.28125
19/64	0.296875
5/16	0.3125

Note: To find the decimal equivalent of a fraction, divide the numerator by the denominator.

FIGURE C.51 Decimal equivalents of fractions of an inch.

APPENDIX

Other useful information

Type of Caulk	Relative Cost	Life Years
Oil	Low	1 to 3
Vinyl latex	Low	3 to 5
Acrylic latex	Low	5 to 10
Silicon acrylic latex	Medium	10 to 20
Butyl rubber	Medium	5 to 10
Polysulfide	Medium	20+
Polyurethane	High	20+
Silicone	High	20+
Urethane foam	High	10 to 20

FIGURE D.1 Characteristics of caulks.

- Gloss finishes are smoother and easier to clean.
- Gloss finishes are usually confined to use on exterior trim, while a flat finish is used on siding.
- The paint on a home exterior should last from 5 to 10 years.
- Exterior painting should not be done once the air temperature drops below 50°F. Extremely hot days are also days to avoid painting.
- Aluminum siding can be painted when the proper paint is used.
- With the exception of slate roofs and glazed tiles, all exterior surfaces are able to accept paint.
- Poor prep work and moisture are the two most common causes of peeling paint.
- Painting surfaces must be completely dry when working with oil-based materials.
- Most professionals paint windows after siding has been painted.
- Gutters and downspouts should be painted the same color as the siding is painted.

FIGURE D.2 Professional suggestions for exterior painting.

doi: 10.1016/B978-1-85617-549-4.00025-3

Tile	Rating
Ceramic tile	Fairly easy
Ceramic mosaic tile	Easy
Quarry tile	Fairly difficult

FIGURE D.3 Difficulty rating for installing tile.

- *Organic adhesive*: This product comes in a ready-mix mastic, applies easily, creates a flexible bond, and is available at low cost. Use of this material is limited to interior applications.
- *Dry-set mortar*: Created with a dry mix of cement, sand, and additives, this material resists freezing and immersion. However, the material must be kept moist for about 3 days prior to grouting.
- *Portland cement mortar*: Portland cement, sand, and water are mixed to create this setting agent. An advantage to this material is that it allows for minor adjustments in leveling the work area. However, tiles must be presoaked, and metal lath reinforcement is recommended.
- *Expoxy mortar*: Mixed by parts on the site, Expoxy mortar creates an extremely strong bond and is highly resistant to water and chemicals. Due to the nature of Expoxy, the time available for working with the quick-setting material is limited.

FIGURE D.4 Suggestions for tile setting materials.

Type of Tile	Size of Tile	Maximum Joint Width	Minimum Joint Width
Ceramic mosaic	2 3/16" or less	1/8 inch	1/16 inch
Ceramic	2 3/16" to 4 1/4"	1/4 inch	1/8 inch
Ceramic	6" × 6"	3/4 inch	1/4 inch
Quarry	All sizes	3/4 inch	3/8 inch

FIGURE D.5 Recommended joint widths for floor tile.

Component	Material	Load (PSF)
Roofing	Softwood (per inch)	3
	Plywood (per inch)	3
	Foam insulation (per inch)	.2
	Asphalt shingle	3
	Asphalt roll roofing	1
	Asphalt (built up)	6
	Wood shingle	3
	Copper	1
	Steel	2
	Slate (3/8")	12
	Roman tile	12
	Spanish tile	19

FIGURE D.6 Weights of building materials for roofing.

Component	Material	Load (PSF)
Framing (16" oc)	2 × 4 and 2 × 6	2
	2 × 8 and 2 × 10	3
Floor-ceiling	Softwood (per inch)	3
	Hardwood (per inch)	4
	Plywood (per inch)	3
	Concrete (per inch)	12
	Stone (per inch)	13
	Carpet	.5
	Drywall (per inch)	5

FIGURE D.7 Weights of building materials for framing and floor.

Thickness	Width	Height
	Exterior	
1¾"	2'8" to 3'0"	6'8" residential
		7'0" commercial
	Interior	
1⅜"	2'6" min. Bedroom	6'8"
	2'0" min. bath, closet	
Hardware		**Placement**
Door knob		36" above floor
Door hinges		11" above floor and 7" down from top of door Optional third hinge ½ way between other 2
		1/16" at top and latch side
		1/32" at hinge side
		5/8" at bottom
Door clearance (Interior doors)		

FIGURE D.8 Typical dimensions of doors.

Pipe Size	Projected Flow Rate (Gallons per Minute)
½ inch	2 to 5
¾ inch	5 to 10
1 inch	10 to 20
1¼ inch	20 to 30
1½ inch	30 to 40

FIGURE D.9 Projected flow rates for various pipe sizes.

Pipe Size	Number of Gallons
¾ inch	2.8
1 inch	4.5
1¼ inch	7.8
1½ inch	11.5
2 inch	18

FIGURE D.10 Fluid volume of pipe contents for polybutylene pipe (computed on the number of gallons per 100 feet of pipe).

Pipe Size	Number of Gallons
1 inch	4.1
1¼ inch	6.4
1½ inch	9.2

FIGURE D.11 Fluid volume of pipe contents for copper pipe (computed on the number of gallons per 100 feet of pipe).

- Windows
- Doors
- Outside walls
- Partitions between heated and unheated space
- Ceilings
- Roofs
- Uninsulated wood floors between heated and unheated space
- Air infiltration through cracks in construction
- People in the building
- Lights in the building
- Appliances and equipment in the building

FIGURE D.12 Common sources of heat gain in buildings.

Material	Heat Storage (Highest Numbers Are Best)
Water	9
Wood	8
Oil	7
Air	6
Aluminum	5
Concrete	4
Glass	4
Steel	3
Lead	2

FIGURE D.13 Heat storage comparisons.

- Windows
- Doors
- Outside walls
- Partitions between heated and unheated space
- Ceilings
- Roofs
- Concrete floors
- Uninsulated wood floors between heated and unheated space
- Air infiltration through cracks in construction

FIGURE D.14 Common sources of heat loss in buildings.

- Bow
- Cup
- Crook
- Twist
- Check
- Split
- Shake
- Wane
- Knot
- Cross grain
- Decay
- Pitch pocket

FIGURE D.15 Types of lumber defects.

Application	Size	Notes
Siding	3/8 inch	
Wall sheathing	5/16 inch	Studs 16" on center
Wall sheathing	3/8 inch	Studs 24" on center
Roof sheathing	5/16 inch	Rafters 16" on center
Roof sheathing	3/8 inch	Rafters 24" on center

FIGURE D.16 Exterior use of plywood.

Usage	Live Load (lbs./sq. ft.)
Bedrooms	30
Other rooms (residential)	40
Ceiling joists (no attic use)	5
Ceiling joists (light storage)	20
Ceiling joists (attic rooms)	30
Retail stores	75 to 100
Warehouses	125 to 250
School classrooms	40
Offices	80
Libraries	
Reading rooms	60
Book stacks	150
Auditoriums, gyms	100
Theater stage	150
Most corridors, lobbies, stairs exits, fire escapes, etc. in public buildings	100

FIGURE D.17 Design loads.

Nominal Size	Actual Size	Board Feet per Linear Foot	Linear Feet per 1,000 Board Feet
1 × 2	¾ × ½	1/16 (.167)	6000
1 × 3	¾ × 2½	¼ (.250)	4000
1 × 4	¾ × 3½	⅓ (.333)	3000
1 × 6	¾ × 5½	½ (.500)	2000
1 × 8	¾ × 7¼	⅔ (.666)	1500
1 × 10	¾ × 9¼	⅚ (.833)	1200
1 × 12	¾ × 11¼	1 (1.0)	1000

FIGURE D.18 Board lumber measure.

Material	Weight in Pounds per Cubic Inch	Weight in Pounds per Cubic Foot
Aluminum	.093	160
Antimony	.2422	418
Brass	.303	524
Bronze	.320	552
Chromium	.2348	406
Copper	.323	558
Gold	.6975	1205
Iron (cast)	.260	450
Iron (wrought)	.2834	490
Lead	.4105	710
Manganese	.2679	463
Mercury	.491	849
Molybdenum	.309	534
Monel	.318	550
Platinum	.818	1413
Steel (mild)	.2816	490
Steel (stainless)	.277	484
Tin	.265	459
Titanium	.1278	221
Zinc	.258	446

FIGURE D.19 Weights of various materials.

Metal	Degrees F
Aluminum	1200
Antimony	1150
Bismuth	500
Brass	1700/1850
Copper	1940
Cadmium	610
Iron (cast)	2300
Iron (wrought)	2900
Lead	620
Mercury	139
Steel	2500
Tin	446
Zinc (cast)	785

FIGURE D.20 Melting points of commercial metals.

Surface	Minimum	Maximum
Driveways in the north	1%	10%
Driveways in the south	1%	15%
Walks	1%	4%
Ramps		15%
Wheelchair ramps		8%
Patios	1%	2%

FIGURE D.21 Grades for traffic surface.

Material	Chemical Symbol
Aluminum	AL
Antimony	Sb
Brass	..
Bronze	..
Chromium	Cr
Copper	Cu
Gold	Au
Iron (cast)	Fe
Iron (wrought)	Fe
Lead	Pb
Manganese	Mn
Mercury	Hg
Molybdenum	Mo
Monel	..
Platinum	Pt
Steel (mild)	Fe
Steel (stainless)	..
Tin	Sn
Titanium	Ti
Zinc	Zn

FIGURE D.22 Symbols for various materials.

Material	Expected Life Span
Aluminum	15 to 20 years
Vinyl	Indefinite
Steel	Less than 10 years
Copper	50 years
Wood	10 to 15 years

Note: All estimated life spans depend on installation procedure, maintenance, and climatic conditions.

FIGURE D.23 Potential life spans for gutters.

Material	Price Range
Aluminum	Moderate
Vinyl	Expensive
Steel	Inexpensive
Copper	Very expensive
Wood	Moderate to expensive

Note: All estimated life spans depend on installation procedure, maintenance, and climatic conditions.

FIGURE D.24 Price ranges for gutters.

GPM	Liters/Minute
1	3.75
2	6.50
3	11.25
4	15.00
5	18.75
6	22.50
7	26.25
8	3.00
9	33.75
10	37.50

FIGURE D.25 Flow rate conversion from gallons per minute (GPM) to approximate liters per minute.

Vacuum in Inches of Mercury	Boiling Point
29	76.62
28	99.93
27	114.22
26	124.77
25	133.22
24	14.31
23	146.45
22	151.87
21	156.75
20	161.19
19	165.24
18	169.00
17	172.51
16	175.80
15	178.91
14	181.82
13	184.61
12	187.21
11	189.75
10	192.19
9	194.50
8	196.73
7	198.87
6	20.96
7	198.87
6	20.96
5	202.25
4	204.85
3	206.70
2	208.50
1	21.25

FIGURE D.26 Boiling points of water at various pressures.

Activity	Normal Use	Conservative Use
Shower	25 (watering running)	4 (wet down, soap up, rinse off)
Tub bath	36 (full)	10 to 12 (minimal water level)
Dishwashing	50 (tap running)	5 (wash and rinse in sink)
Toilet flushing	5 to 7 (depends on tank size)	1½ to 3 (Water-Saver toilets or tank displacement bottles)
Automatic dishwasher	16 (full cycle)	7 (short cycle)
Washing machine	60 (full cycle, top water level)	27 (short cycle, minimal water level)
Washing hands	2 (tap running)	1 (full basin)
Brushing teeth	1 (tap running)	½ (wet and rinse briefly)

FIGURE D.27 Conserving water usage in gallons.

- A cubic foot of water contains 7½ gallons, 1728 cubic inches, and weights 62½ pounds.
- A gallon of water weighs 8⅓ pounds and contains 231 cubic inches.
- Water expands $^{1}/_{23}$ of its volume when heated from 40° to 212°.
- The height of a column of water, equal to a pressure of 1 pound per square inch, is 2.31 feet.
- To find the pressure in pounds per square inch of a column of water, multiply the height of the column in feet by .434.
- The average pressure of the atmosphere is estimated at 14.7 pounds per square inch so that with a perfect vacuum it will sustain a column of water 34 feet high.
- The friction of water in pipes varies as the square of the velocity.
- To evaporate 1 cubic foot of water requires the consumption of 7½ pounds of ordinary coal or about 1 pound of coal to 1 gallon of water.
- A cubic inch of water evaporated at atmospheric pressure is converted into approximately 1 cubic foot of steam.

FIGURE D.28 Facts about water.

Square inch = 0.007 square feet = 6.45 square centimeters

Square foot = 144 square inches = 929.03 square centimeters

Square yard = 9 square feet = 0.836 square meters

Square rod = 30.25 square yards
Acre = 4,840 square yards = 43,560 square feet = 160 square rods = 4.047 square meters = 0.405 hectare

Hectare = 10,000 square meters = 2.47 acres

Square mile = 640 acres = 2.59 sq kilometers = 1 section

Section = 1 square mile = 640 acres = 2.59 square kilometers

FIGURE D.29 Measurement conversion table.

Joint Size Depth × Width (Inches)	Lineal Feet per U.S. Gallon	Joint Size Depth × Width (Millimeters)	Meters per Liter
1/8 × 1/8	1232.0	3.2 × 3.2	99.20
1/8 × 1/4	616.0	3.2 × 6.4	49.60
1/8 × 3/8	410.7	3.2 × 9.5	33.07
1/8 × 1/2	308.0	3.2 × 12.7	24.80
1/8 × 5/8	246.4	3.2 × 15.9	19.84
1/8 × 3/4	205.3	3.2 × 19.1	16.53
1/8 × 7/8	176.0	3.2 × 22.2	14.17
1/8 × 1	154.0	3.2 × 25.4	12.40
1/4 × 1/4	308.0	6.4 × 6.4	24.80
1/4 × 3/8	205.0	6.4 × 9.5	16.53
1/4 × 1/2	154.0	6.4 × 12.7	12.40
1/4 × 5/8	123.2	6.4 × 15.9	9.92
1/4 × 3/4	102.7	6.4 × 19.1	8.27
1/4 × 7/8	88.0	6.4 × 22.2	7.09
1/4 × 1	77.0	6.4 × 25.4	6.20
3/8 × 3/8	136.9	9.5 × 9.5	11.02
3/8 × 1/2	102.7	9.5 × 12.7	8.27
3/8 × 5/8	82.1	9.5 × 15.9	6.61
3/8 × 3/4	68.4	9.5 × 19.1	5.51
3/8 × 7/8	58.7	9.5 × 22.2	4.72
3/8 × 1	51.3	9.5 × 25.4	4.13
1/2 × 1/2	77.0	12.7 × 12.7	6.20
1/2 × 5/8	61.6	12.7 × 15.9	4.96
1/2 × 3/4	51.3	12.7 × 19.1	4.13
1/2 × 7/8	44.0	12.7 × 22.2	3.54
1/2 × 1	38.5	12.7 × 25.4	3.10
5/8 × 5/8	49.3	15.9 × 15.9	3.97
5/8 × 3/4	41.1	15.9 × 19.1	3.31
5/8 × 7/8	35.2	15.9 × 22.2	2.83
5/8 × 1	30.8	15.9 × 25.4	2.48

FIGURE D.30 Caulking coverage.

Aggregate Designation	50 mm (2 in.)	37.5 mm (1 ½ in.)	31.5 mm (1 ¼ in.)	25.0 mm (1 in.)	19.0 mm (¾ in.)	16.0 mm (⅝ in.)	12.5 mm (½ in.)	9.5 mm (⅜ in.)	4.75 mm (# 4)
CA–00		100			95–100		0–10		
CA–15	100		90–100		35–65	5–25		0–7	
CA–25 or 2M6	100		95–100		50–80	20–40		0–7	
CA–35 or 3M6	100		95–100		55–85	20–45		0–7	
CA–45 or 4M6	100		95–100		65–95	25–55		0–7	
CA–50	100		85–100			30–60		0–12	
CA–60	100		85–100			40–70		0–12	
CA–70	100		85–100			50–100		0–25	
CA–80 *(A)*			100			55–95			

FIGURE D.31 Coarse aggregate designation for concrete (percent by mass (weight) passing square opening sieves *(A)*).

Problems	Potential Result(s)	Possible Cause(s)	What to Do
False set	Stiff, unworkable mixture	Form of gypsum in cement and/or admixture incompatibility	Perform additional mixing
Flash set	Reduced workability, poor rapid set (cannot be fixed)	Low gypsum content in Portland cement	Get new supply of Portland cement
Equipment breakdown	Costly reductions in productivity and pavement quality	Poor maintenance	Maintain equipment regularly
High water/cement (w/c) ratio	Reduced strength	Adding water on site	Adjust the amount of water added at plant; do not add water at the construction site
Inadequate consolidation and workability	Reduced strength and durability	Vibrator problems; inadequate mixing	Monitor vibrators for compliance and repair as needed; provide adequate mixing
Inadequate entrained air	Reduced concrete durability; possible negative effect on pay factors	Weather, short mixing time	Monitor consistently, especially on extreme weather days
Dips in pavement profile	Reduced pavement smoothness; Variations in slab thickness	Moved or otherwise disturbed stringline; ruts or irregularities in subbase surface	Place stringline as low as possible; monitor stringline regularly and notify supervisor of stringline disturbances; ensure smooth, rut-free subbase surface
Bumpy, rutty haul road	Increased mix delivery time; reduced productivity; possible pumping of stringline pins	Poorly maintained haul road	Maintain road during construction
Sudden weather change: rain	Increased w/c ratio on the surface; reduced durability on the surface; loss of texture		Stop paving; cover the slab to protect it from rain damage
Sudden weather change: cold front	Stresses due to sudden temperature change that can result in increased random cracking		Protect fresh concrete with additional curing; consider insulation

FIGURE D.32 Common concrete problems and solutions.

Common Volume Conversion Factors	
1 cubic centimeter =	1000 cubic millimeter
1 cubic decimeter =	1000 cubic centimeter
1 cubic meter =	1000 cubic decimeter
1 liter/litre =	.001 cubic meter
1 liter/litre =	10 deciliter
1 deciliter =	10 centiliter
1 centiliter =	10 milliliter
1 cubic foot =	1728 cubic inches
1 cubic yard =	27 cubic feet

U.K. Capacity Conversion Factors	
1 fluid minims =	59.19379 cu mm
1 fluid drams =	60 minims
1 fluid ounces =	8 fluid drams
1 pint (pt) =	20 fluid ounces
1 gills (gi) =	0.25 pint
1 quart (qt) =	2 pint
1 gallon (gal) =	4 quart
1 oil barrel =	36 gallons
1 peck (pk) =	8 quart
1 bushel (bu) =	32 quart

U.S. Liquid Volume Conversion Factors	
1 fluid minims =	61.61152 cu mm
1 fluid drams =	60 minims
1 fluid ounces =	8 fluid drams
1 pint (pt) =	16 fluid ounces
1 gills (gi) =	0.25 pint
1 quart (qt) =	2 pint
1 gallon (gal) =	4 quart
1 oil barrel =	42 gallons

FIGURE D.33 Conversions.

U.S. Dry Volume Conversion Factors	
1 pint (pt) =	33.6 cubic inches
1 quart (qt) =	2 pint
1 gallon (gal) =	4 quart
1 peck (pk) =	8 dry quart
1 bushel (bu) =	32 dry quart

Other Volume Conversion Factors	
1 cord (cd) =	128 cubic feet
1 Teaspoon =	5 Millilitre
1 Tablespoon =	3 Teaspoon
1 Cup =	16 Tablespoon

FIGURE D.33 *(Continued)*

Product Name	Manufacturer	Manufacturer Website	Type	Min. Diameter, Min. Embedment Depth, & Notes
Spec-Bond 201	Conspec	*http://www.conspecmkt.com*	Adhesive/Epoxy	7/8" dia., 3-1/2" length
Sure-Grip	Dayton Superior	*http://www.daytonsuperior.com*	Adhesive/Epoxy	
Sure Anchor J-51	Dayton Superior	*http://www.daytonsuperior.com*	Adhesive/Epoxy	1/2" dia., 4.5" embedment
Wedge Anchor	DFS Inc.		Mechanical Anchor	7/8" dia., 3-1/2" length
Dyna Grout	Dyna Grout	*http://www.dynagrout.com*	Grout	
HY-150	Hilti Fastening	*http://www.hilti.com*	Adhesive/Acrylic	1/2" dia., 4.25" embedment
HVA Adhesive System	Hilti Fastening	*http://www.hilti.com*	Vinyl Urethane-Adhesive System	5/8" dia., 5" embedment
Kwik Bolt	Hilti Fastening	*http://www.hilti.com*	Mechanical Anchor	1/2" dia., 4" embedment
HIT RE500	Hilti Fastening	*http://www.hilti.com*	Adhesive/Epoxy	3/8" dia., 3-3/8" embedment
HIT-ICE	Hilti Fastening	*http://www.hilti.com*	Adhesive System	3/8" dia., 2" embed/low temp
HIT HY 150 MAX	Hilti Fastening	*http://www.hilti.com*	Adhesive System	3/8" dia., 2-1/4" embedment
RSE DOT	Hilti Fastening	*http://www.hilti.com*	Adhesive System	3/8" dia., 3 3/8" embedment
Red Head Anchors	ITT Philipps Drill	*http://www.rsci.com/tools*	Mechanical Anchor	1/2" dia., 4" embedment
Epcon Granite 5	ITW Ramset/Redhead	*http://www.ramset-redhead.com*	Adhesive/Epoxy	3/8" dia., 3 3/8" embedment
Epcon System Acrylic 7	ITW Ramset/Redhead	*http://www.ramset-redhead.com/*	Adhesive/Acrylic	1/2" dia., 4.5" embedment
Keligrout	Kelken Construction Systems, Inc.	*http://www.kelken.com*	Adhesive/Resin	1/2" dia., 5"embed/low temp

FIGURE D.34 Concrete anchorage.

Product Name	Manufacturer	Manufacturer Website	Type	Min. Diameter, Min. Embedment Depth, & Notes
Keligrout 101-P	Kelken Construction Systems, Inc.	*http://www.kelken.com*	Adhesive/Resin	1/2" dia., 5" embed/low temp
Liebig Safety Bolts	Liebig International	*http://www.liebig-profibolt.com*	Mechanical Anchor	1/2" dia., 4" embedment
Liquid Roc 300 Capsules	MKT Fastening	*http://www.mktfastening.com*	Polyester Capsule	3/8" dia., 3-3/8" embedment
Liquid Roc 500 Low Odor	MKT Fastening	*http://www.mktfastening.com*	Adhesive/Epoxy	3/8" dia., 3-3/8" embedment
Polybac 1257	Polygem	*http://www.polygem.com*	Adhesive/Epoxy	1/2" dia., 4.5" embedment
Power-Fast Standard Set	Powers Fasteners, Inc.	*http://www.powers.com*	Adhesive/Epoxy	1/2" dia., 4.5" embedment
AC 100 + Gold	Powers Fasteners, Inc.	*http://www.powers.com*	Adhesive System	5/8" dia., 5-5/8" embedment
Power-Stud	Powers Fasteners, Inc.	*http://www.powers.com*	Mechanical Anchor	1/2" dia., 4" embedment
Speed Bond #1	Prime Resins	*http://www.primeresins.com*	Adhesive/Epoxy	5/8" dia., 6 1/4" embedment
Fastlane	Prime Resins	*http://www.primeresins.com*	Adhesive/Epoxy	5/8" dia., 6 1/4" embedment
Sikadur DOT-SP13	Sika Corporation	*http://www.sikaconstruction.com*	Adhesive/Epoxy	1/2" dia., 5" embedment
Sikadur Fast Injection Gel	Sika Corporation	*http://www.sikaconstruction.com*	Adhesive/Epoxy	1/2" dia., 5" embedment
Epoxy Tie (ET)	Simpson Strong-Tie	*http://www.strongtie.com*	Adhesive/Epoxy	5/8" dia., 5" embedment
Epoxy Tie Fast Cure (ETF)	Simpson Strong-Tie	*http://www.strongtie.com*	Adhesive/Epoxy	5/8" dia., 5" embedment

(Continued)

Product Name	Manufacturer	Manufacturer Website	Type	Min. Diameter, Min. Embedment Depth, & Notes
Set Adhesive (SET)	Simpson Strong-Tie	*http://www.strongtie.com*	Adhesive/Epoxy	5/8" dia., 5" embedment
Acrylic Tie (AT)	Simpson Strong-Tie	*http://www.strongtie.com*	Adhesive/Acrylic	3/8" dia., 3.5" embedment
Wedge-All Expansion Anchor	Simpson Strong-Tie	*http://www.strongtie.com*	Mechanical Anchor	3/4" dia., 3-3/8" embedment
EDOT	Simpson Strong-Tie	*http://www.strongtie.com/*	Adhesive/Epoxy	3/8" dia., 3-3/8' embedment
Pro-Poxy 300	Unitex	*http://www.unitexchemical.com*	Adhesive/Epoxy	1/2" dia., 4.5" embedment
Pro-Poxy 300 Fast	Unitex	*http://www.unitexchemical.com*	Adhesive/Epoxy	1/2" dia., 4.5" embedment
Thunderbolts	Universal Fastenings	*http://www.ufss.com*	Mechanical Anchor	1/2" dia., 4" embedment
HS-200	US Anchor Corp.	*http://www.usanchor.com*	Adhesive/Epoxy	3/8" dia., 3-1/2" embedment Use when temp. below 60°F., & allow 48 hours before loads.
REZI-WELD Gel Paste	W.R. Meadows, Inc.	*http://www.wrmeadows.com*	Adhesive/Epoxy	3/4" dia., 7-1/2" embedment

Ton	Inch	Square Feet	Square Yards
1 ton	1.0 inch by	249.23	27.69
1 ton	2.0 inches by	124.62	13.85
1 ton	3.0 inches by	83.08	9.23
1 ton	4.0 inches by	62.31	6.92
1 ton	5.0 inches by	49.85	5.54
1 ton	6.0 inches by	41.54	4.62
1 ton	7.0 inches by	35.60	3.96

FIGURE D.35 Concrete sand estimations.

ADVERTISEMENT FOR BIDS
The public announcement, as required by law, inviting bids for the work to be performed or materials to be furnished.

AGGREGATE
Natural materials such as sand, gravel, crushed rock, or taconite tailings, and crushed concrete or salvaged bituminous mixtures, usually with a specified particle size, for use in base course construction, paving mixtures, and other specified applications.

AUXILIARY LANE
The portion of the roadway adjoining the traveled way for parking, speed-change, or other purposes supplementary to through traffic movement.

AWARD
The acceptance by the Contracting Authority of a bid, subject to execution and approval of the Contract.

BID SCHEDULE
A listing of Contract items in the proposal form showing quantities and units of measurement that provides for the bidder to insert unit bid prices.

BIDDER
An individual, firm, or corporation submitting a Proposal for the advertised work.

BRIDGE
A structure, including supports, erected over a depression or an obstruction, such as a water course, highway, or railway, and having a track or passageway for carrying traffic or other moving loads. Traffic or other moving loads are carried directly on the upper portion of the superstructure (called the bridge deck).

BRUSH
Shrubs, trees, and other plant life having a diameter of 100 mm (**4 inches**) or less at a point 600 mm (**12 inches**) above ground surface, as well as fallen trees and branches.

CALENDAR DAY
Every day shown on the calendar.

CARBONATE
Sedimentary rock composed primarily of carbonate minerals, including dolostone (dolomite, $CaMg(CO_3)_2$), limestone (calcite, $CaCO_3$) and mixtures of dolostone and limestone.

CERTIFICATE OF COMPLIANCE
A certification provided by a manufacturer, producer, or supplier of a product that the product, as furnished to the Contractor, complies with the pertinent specification or Contract requirements. The certification shall be signed by a person who is authorized to bind the company supplying the material covered by the certification.

CERTIFIED TEST REPORT
A test report provided by a manufacturer, producer, or supplier of a product indicating actual results of tests or analyses, covering elements of the specification requirements for the product or workmanship, and including validated certification.

CERTIFIED CCTV TECHNICIAN
An individual certified by the Contractor and approved by the Engineer to perform all work associated with a CCTV system.

CHANGE ORDER
A written order issued by the Engineer to the Contractor covering permissible adjustments, minor Plan changes or corrections, and rulings with respect to omissions, discrepancies, and intent of the Plans and Specifications, but not including any Extra Work or other alterations that are required to be covered by Supplemental Agreement.
Orders issued to implement changes made by mutual agreement shall not become effective until signed by the Contractor and returned to the Engineer.

CHANGED CONDITION
The Contract clause (1402) that provides for adjustment of Contract terms for site conditions that differ from those in the Contract, for suspension of work ordered by the Department, and for significant changes in the character of the work.

CITY, VILLAGE, TOWNSHIP, TOWN, OR BOROUGH
A subdivision of the county used to designate or identify the location of the proposed work. **1103** 7.

COMMISSIONER
The Commissioner of the Minnesota Department of Transportation, or the chief executive of the department or agency constituted for administration of Contract work within its jurisdiction.

CONTRACT
The written agreement between the Contracting Authority and the Contractor setting forth their obligations, including, but not limited to, the performance of the work, the furnishing of labor and materials, the basis of payment, and other requirements contained in the Contract documents.
The Contract documents include the advertisement for bids, Proposal, Contract form, Contract bond, these Specifications, supplemental Specifications, Special Provisions, general and detailed Plans, notice to proceed, and orders and agreements that are required to complete the construction of the work in an acceptable manner, including authorized extensions, all of which constitute one instrument.

CONTRACT BOND
The approved form of security executed by the Contractor and Surety or Sureties, guaranteeing complete execution of the Contract and all Supplemental Agreements pertaining thereto and the payment of all legal debts pertaining to construction of the Project.

CONTRACT ITEM (Pay Item)
A specifically described unit of work for which a price is provided for in the Contract.

CONTRACT TIME
The completion date, number of working days, or number of calendar days allowed for completion of the Contract, including authorized time extensions.
Completion date and calendar day Contracts shall be completed on or before the day indicated even where that date is a Saturday, Sunday, or holiday.

CONTRACTING AUTHORITY
The political subdivision, governmental body, board, department, commission, or officer making the award and execution of Contract as the party of the first part.

CONTRACTOR
The individual, firm, or corporation contracting for and undertaking prosecution of the prescribed work; the party of the second part to the Contract, acting directly or through a duly authorized representative.

COUNTY
The county in which the prescribed work is to be done; a subdivision of the State, acting through its duly elected Board of County Commissioners.

CULVERT
A structure constructed entirely below the elevation of the roadway surface and not a part of the roadway surface, which provides an opening under the roadway for the passage of water or traffic.

DEPARTMENT
The Department of Transportation of the State of Minnesota, or the political subdivision, governmental body, board, commission, office, department, division, or agency constituted for administration of the Contract work within its jurisdiction.

DETOUR
A road or system of roads, usually existing, designated as a temporary route by the Contracting Authority to divert through traffic from a section of roadway being improved.

DIVIDED HIGHWAY
A highway with separated traveled ways for traffic in opposite directions.

DORMANT SEEDING
Seeding allowed in the late fall when the ground temperature is too low to cause seed germination so the seed remains in a dormant condition until spring.

DORMANT SODDING
Sodding allowed in the late fall when the ground temperature is too low so that normal rooting does not take place until spring.

EASEMENT
A right acquired by public authority to use or control property for a designated highway purpose.

ENGINEER
The duly authorized engineering representative of the Contracting Authority, acting directly or through the designated representatives who have been delegated responsibility for engineering supervision of the construction, each acting within the delegated scope of duties and authority.

EQUIPMENT
All machinery, tools, and apparatus, together with the necessary supplies for upkeep and maintenance, necessary for the proper construction and acceptable completion of the Contract within its intended scope.

EROSION CONTROL SCHEDULE
An oral commitment or written document by the Contractor illustrating construction sequences and proposed methods to control erosion.

EXTRA WORK
Any work not required by the Contract as awarded, but which is authorized and performed by Supplemental Agreement, either at negotiated prices or on a Force Account basis as provided elsewhere in these Specifications.

FRONTAGE ROAD (OR STREET)
A local road or street auxiliary to and located on the side of a highway for service to abutting property and adjacent areas and for control of access.

GRADE SEPARATION
A bridge with its approaches that provides for highway or pedestrian traffic to pass without interruption over or under a railway, highway, road, or street.

GRAVEL
Naturally occurring rock or mineral particles produced by glacial and water action. Particle size ranges from 76 mm (**3 inches**) diameter to the size retained on a 2.0 mm diameter (# **10 sieve**).

GUARANTEED ANALYSIS
A guarantee from a manufacturer, producer, or supplier of a product that the product complies with the ingredients or specifications indicated on the product label.

HIGHWAY, STREET, OR ROAD
A general term denoting a public way for purposes of vehicular travel, including the entire area within the right of way.

HOLIDAYS
The days of each year set aside by legal authority for public commemoration of special events, and on which no public business shall be transacted except as specifically provided in cases of necessity. Unless otherwise noted, holidays shall be as established in MS 645.44.

INCIDENTAL
Whenever the word "incidental" is used in the plan or special provisions it shall mean no direct compensation will be made.

INDUSTRY STANDARD
An acknowledged and acceptable measure of quantitative or qualitative value or an established procedure to be followed for a given operation within the given industry. This will generally be in the form of a written code, standard, or specification by a creditable association.

INSPECTOR
The Engineer's authorized representative assigned to make detailed inspections of Contract performance.

INTERCHANGE
A grade-separated intersection with one or more turning roadways for travel between intersection legs.

INTERSECTION
The general area where two or more highways join or cross, within which are included the roadway and roadside facilities for traffic movements in the area.

LIMESTONE
See Carbonate

LOOP
A one-way turning roadway that curves about 270 degrees to the right, primarily to accommodate a left-turning movement, but which may also include provisions for another turning movement.

MATERIALS
Any substances specified for use in the construction of the Project and its appurtenances.

MATERIALS LABORATORY
The Mn/DOT Central Materials Laboratory and, for those tests so authorized, a Mn/DOT District Materials Laboratory.

MAXIMUM DENSITY
The maximum density of a particular soil as determined by the method prescribed in the Mn/DOT Grading and Base Manual.

METRIC
The International System of Units (SI) (the Modernized Metric System) according to ASTM E 380 are used in these Specifications. ASTM E 380 also provides conversion factors and commentary.

METRIC TON (t)
A mass of 1000 kg.

MINOR EXTRA WORK
Extra Work ordered by the Engineer and required to complete the project as originally intended and authorized in writing.

MGal
1000 Gallons

NOMINAL
The intended, named, or stated value, as opposed to the actual value. The nominal value of something is the value that it is supposed or intended to have, or the value by which it is commonly known. The actual value may differ from these statements by a greater or lesser amount depending on the accuracy and precision of the process used to determine the actual value.

NOTICE TO PROCEED
Written notice to the Contractor to proceed with the Contract work including, when applicable, the date of beginning of Contract time.

NPDES PERMIT
The general permit issued by the MPCA that authorized the discharge of storm water associated with construction activity under the National Pollutant Discharge Elimination System Program.

OPTIMUM MOISTURE
The moisture content of a particular soil at maximum dry density as determined by the method prescribed in the Mn/DOT Grading and Base Manual.

(P)
A designation in the summary of quantities in the Plan, meaning that the Plan quantity will be the quantity for payment. Measurement or recomputation will not be made except as provided in 1901.

PAY ITEM (Contract Item)
A specifically described unit of work for which a price is provided for in the Contract.

PAVEMENT STRUCTURE
The combination of subbase, base course, and surface course placed on a subgrade to support the traffic load and distribute it to the roadbed.

PERMANENT EROSION CONTROL MEASURES
Soil-erosion control measures such as curbing, culvert aprons, riprap, flumes, sodding, erosion mats, and other means to permanently minimize erosion on the completed Project.

PLAN
The Plan, profiles, typical cross-sections, and supplemental drawings that show the locations, character, dimensions, and details of the work to be done.

PLAN QUANTITY
The quantity listed in the summary of quantities in the Plan. The summary of quantities will usually be titled Statement of Estimated Quantities, Schedule of Quantities for Entire Bridge, or Schedule of Quantities.

PROFILE GRADE
The trace of a vertical plane intersecting the top surface of the roadbed or pavement structure, usually along the longitudinal centerline of the traveled way. Profile grade means either elevation or gradient of such trace according to the context.

PROJECT
The specific section of the highway, the location, or the type of work together with all appurtenances and construction to be performed under the Contract.

PROPOSAL
The offer of a bidder on the prescribed Proposal form to perform the work and furnish the labor and materials at the prices quoted.

PROPOSAL FORM
The approved form on which the Contracting Authority requires bids to be prepared and submitted for the work.

PROPOSAL GUARANTY
The security furnished with a bid to guarantee that the bidder will enter into the Contract if the bid is accepted.

PURE LIVE SEED (Percentage)
A percentage determined by the percent of seed germination times the percent of seed purity.

QUALITY ASSURANCE (QA)
The activities performed by the Department that have to do with making sure the quality of a product is what it should be.

QUALITY COMPACTION
A compaction method as defined in 2105.3F2.

QUALITY CONTROL (QC)
The activities performed by the Contractor that have to do with making the quality of a product what it should be.

QUESTIONNAIRE
The specified forms on which a bidder may be required to furnish information as to ability to perform and finance the work.

RAMP
A connecting roadway for travel between intersection legs at or leading to an interchange.

RIGHT OF WAY
A general term denoting land, property, or interest therein, usually in a strip, acquired for or devoted to a highway.

ROAD
A general term denoting a public way for purposes of vehicular travel, including the entire area within the right of way.

ROADBED
The graded portion of a highway within top and side slopes, prepared as a foundation for the pavement structure and shoulders.

ROADWAY
The portion of a highway within limits of construction.

SCALE
A device used to measure the mass or the proportion of a liquid or solid. This definition includes metering devices.

SHOULDER
The portion of the roadway contiguous with the traveled way for accommodation of stopped vehicles, for emergency use, and for lateral support of the base and surface courses.

SIDEWALK
That portion of the roadway primarily constructed for the use of pedestrians.

SIEVE
A woven wire screen meeting the requirements of AASHTO M-92 for the size specified.

SPECIAL PROVISIONS
Additions and revisions to the standard and supplemental Specifications covering conditions peculiar to an individual Project.

SPECIFICATIONS
A general term applied to all directions, provisions, and requirements pertaining to performance of the work.

SPECIFIED COMPLETION DATE
The date on which the Contract work is specified to be completed.

SPECIMEN TREE
Historic or otherwise significant tree indicated in the Contract or determined by the Engineer.

STATE
The State of Minnesota acting through its elected officials and their authorized representatives.

STREET
A general term denoting a public way for purposes of vehicular travel, including the entire area within the right of way.

STRUCTURES
Bridges, culverts, catch basins, drop inlets, retaining walls, cribbing, manholes, endwalls, buildings, sewers, service pipes, underdrains, foundation drains, and other man-made features.

SUBCONTRACTOR
An individual, firm, or corporation to whom the Contractor sublets part of the Contract.

SUBGRADE
The top surface of a roadbed upon which the pavement structure and shoulders are constructed. Also, a general term denoting the foundation upon which a base course, surface course, or other construction is to be placed, in which case reference to subgrade operations may imply depth as well as top surface.

SUBSTRUCTURE
The part of a bridge below the bearings of simple and continuous spans, skewbacks, or arches and tops of footings for rigid frames, together with the backwalls, wingwalls, and wing protection railings.

SUPERINTENDENT
The Contractor's authorized representative in responsible charge of the work.

SUPERSTRUCTURE
The entire bridge except the substructure.

SUPPLEMENTAL AGREEMENT
A written agreement between the Contracting Authority and the Contractor, executed on the prescribed form and approved as required by law, covering the performance of Extra Work or other alterations or adjustments as provided for within the general scope of the Contract, but which Extra Work or Change Order constitutes a modification of the Contract as originally executed and approved.

SUPPLEMENTAL DRAWINGS
An approved set of drawings consisting of standard plates or plans showing the details of design and construction for various structures and products for which standards have been developed. These standard plates and plans shall govern by reference as identified and supplemented or amended in the general Plans and Specifications.

SUPPLEMENTAL SPECIFICATIONS
Additions and revisions to the standard Specifications that are approved subsequent to issuance of the printed book of standard Specifications.

SURETY
The Corporation, partnership, or individual, other than the Contractor, executing a bond furnished by the Contractor.

TEMPORARY BY-PASS
A section of roadway, usually within existing right of way, provided to temporarily carry all traffic around a specific work site.

TEMPORARY EROSION CONTROL MEASURES
Soil-erosion control measures such as bale checks, silt curtains, sediment traps, and other means to temporarily protect the Project from erosion before and during the installation of permanent erosion control measures. Temporary erosion control measures may also be used to supplement the permanent measures.

TRAFFIC LANE
The portion of a traveled way for the movement of a single line of vehicles.

TRAVELED WAY
The portion of the roadway for the movement of vehicles, exclusive of shoulders and auxiliary lanes.

TURN LANE
An auxiliary lane for left or right turning vehicles.

WORK
The furnishing of all labor, materials, equipment, and other incidentals necessary or convenient to the successful completion of the Project and the carrying out of all the duties and obligations imposed by the Contract upon the Contractor. Also used to indicate the construction required or completed by the Contractor.

WORKING DAY
A calendar day, exclusive of Saturdays, Sundays, and State recognized legal holidays, on which weather and other conditions not under the control of the Contractor will permit construction operations to proceed for at least 2 hours, with the normal working force engaged in performing the progress-controlling operations.

WORKING DRAWINGS
Stress sheets, shop drawings, erection plans, falsework plans, framework plans, cofferdam plans, bending diagrams for reinforcing steel, or any other supplementary plans or similar data that the Contractor is required to furnish and submit to the Engineer.

WORK ORDER
A written order signed by the Engineer of a contractual status requiring performance or other action by the Contractor without negotiation of any sort.

WORK ORDER/MINOR EXTRA WORK
A written order signed by the Engineer requiring performance of Minor Extra Work.

FIGURE D.36 Contract terms and explanations.

Product Name	Manufacturer	Manufacturer Website		
Dayton Superior White PAMS Cure	Dayton Superior	*http://www.daytonsuperiorchemical.com*		
TK Products M-1066	TK Procucts	*http://www.tkproducts.com*		
Vexcon Certi-Vex Enviocure 1315 VOC	Vexcon Chemicals	*http://www.vexcon.com*		
W.R. Meadows Sealtight 2255	W.R. Meadows	*http://www.wrmeadows.com*		

FIGURE D.37 Curing compounds for concrete pavement and bridges.

Product Name	Manufacturer	Manufacturer Website	Preapproved Lots/Batches	Expiration Date
Dayton Superior Resin Cure White	Dayton Superior	*http://www.daytonsuperiorchemical.com*	None	
Right Pointe WW Resin	Right Pointe	*http://www.rightpointe.com*	82921D	04/29/2009
SpecChem Pave Cure Rez White	SpecChem	*http://www.specchemllc.com*	022509PCRW	03/03/2010
			030209PCRW	03/03/2010
			050508PCRW	05/06/2009
			061608PCRW	06/20/2009
Universal Form Clamp White Cure Resin DOT	Universal Form Clamp	*http://www.universalformclamp.com*	None	
Vexcon Certi-Vex Enviocure White 1000	Vexcon Chemicals	*http://www.vexcon.com*	None	
W.R. Meadows Sealtight 1250	W.R. Meadows	*http://www.wrmeadows.com*	M-8HE227	06/10/2009
			M-8HF135	06/26/2009
			M-8HJ103	08/14/2009
			M-8HK062	09/19/2009
			M-9HA049	01/29/2010

FIGURE D.38 Curing compounds.

Diameter	Inches per Lineal Inch	Cubic Feet per Lineal Foot
6	28.27	0.196
8	50.27	0.349
10	78.60	0.550
12	113.10	0.785
16	201.07	1.400
18	254.47	1.767
22	314.00	1.800
24	454.39	3.142
30	706.86	4.910
36	1,017.88	7.070

FIGURE D.39 Cylinder volume conversion.

	Diameter in Inches						
	6	**8**	**10**	**12**	**14**	**16**	**18**
1	0.007	0.013	0.02	0.029	0.039	0.051	0.065
2	0.02	0.03	0.04	0.06	0.08	0.1	0.14
3	0.02	0.04	0.06	0.09	0.12	0.15	0.21
4	0.03	0.05	0.08	0.12	0.16	0.2	0.28
5	0.04	0.07	0.1	0.15	0.2	0.25	0.35
6	0.04	0.08	0.12	0.17	0.23	0.3	0.42
7	0.05	0.09	0.14	0.2	0.27	0.35	0.49
8	0.06	0.1	0.16	0.23	0.32	0.4	0.56
9	0.07	0.12	0.18	0.26	0.36	0.45	0.63
10	0.07	0.13	0.20	0.29	0.4	0.5	0.7
12	0.09	0.16	0.24	0.35	0.48	0.6	0.84
14	0.10	0.18	0.28	0.41	0.56	0.7	0.98
16	0.12	0.21	0.32	0.46	1.64	0.8	1.12
18	0.13	0.23	0.36	0.52	0.72	0.9	1.26
20	0.15	0.26	0.4	0.58	0.8	1	1.4
22	0.16	0.29	0.44	0.64	0.88	1.1	1.54
24	0.18	0.31	0.48	0.7	0.96	1.2	1.68
26	0.19	0.34	0.52	0.75	1.04	1.3	1.82
28	0.20	0.36	0.56	0.81	1.12	1.4	1.96
30	0.22	0.39	0.6	0.87	1.2	1.5	2.1
32	0.23	0.42	0.64	0.93	1.28	1.6	2.24
34	0.25	0.44	0.68	0.99	1.36	1.7	2.38
36	0.26	0.47	0.72	1.04	1.44	1.8	2.52
40	0.29	0.52	0.8	1.16	1.6	2	2.8
	Cubic Yards						

FIGURE D.40 Sonotubes.

Product Name	Manufacturer	Manufacturer Website
Rapid Patch - VR	Bonsal	*http://www.bonsalamerican.com*
Fast Set Cement Mix	Bonsal	*http://www.bonsalamerican.com*
Fast Patch 928	Burke	*http://www.burke.com*
Tectonite	CFB	
Pave Patch 3000	Conspec	*http://www.conspecmkt.com*
Re-Crete 20 Minute Set	Dayton Superior	*http://www.daytonsuperior.com*
Five Star Highway Patch	Five Star Products	*http://www.fivestarproducts.com*
Highway Patch 928	Specco Industries Inc.	*http://www.specco.com*
Speccopatch RS	Specco Industries Inc.	*http://www.specco.com*
Speed Crete 2028	Tamms Industries	*http://www.euclidchemical.com*
Speed Crete Greenline	Tamms Industries	*http://www.euclidchemical.com*
Speed Crete Express Repair	Tamms Industries	*http://www.euclidchemical.com*
Uni Road Repair DOT	Universal Form Clamp	*http://www.universalformclamp.com*
Polypatch	US Mix Products	*http://www.usmix.com*
Transpatch	US Mix Products	*http://www.usmix.com*
Futura	W.R. Meadows	*http://www.wrmeadows.com*
10–60 Rapid Mortar	BASF Construction Chemical	*http://www.construction-chemicals.basf.com/*
10–61 Rapid Mortar	BASF Construction Chemical	*http://www.construction-chemicals.basf.com/*

FIGURE D.41 Fast-hardening concrete mixes for repairs.

ENDURACON TECHNOLOGIES

http://www.enduracon.com

Power Plant	Class	Specific Gravity	Mn/DOT Standard Abbreviation
Dairyland Power - Poz AC @ Alma, WI	C	2.65	POZALWI

ENVIRONMENTAL RESOURCE CORPORATION

Power Plant	Class	Specific Gravity	Mn/DOT Standard Abbreviation
Big Stone @ Big Stone, SD	C	2.65	BSTBSSD

HEADWATERS RESOURCES

http://www.headwaters.com

Power Plant	Class	Specific Gravity	Mn/DOT Standard Abbreviation
Coal Creek @ Underwood, ND	C/F	2.50	COCUNND
Lansing, IA	C	2.75	LANLAIA
Port Neal #3 @ Sioux City, IA	C	2.65	PN3SCIA
Port Neal #4 @ Sioux City, IA	C	2.63	PN4SCIA

LAFARGE NORTH AMERICA

http://www.lafargenorthamerica.com

Power Plant	Class	Specific Gravity	Mn/DOT Standard Abbreviation
Columbia Generating Stations @ Portage, WI	C	2.77	CGSPOWI
Edgewater Unit #5 @ Sheyboygan, WI	C	2.62	EU5SHWI
Northshore Mining @ Silver Bay, MN	C	2.73	NMSBMN
Pleasant Prairie @ Kenosha, WI	C	2.50	PLPKEWI
Weston Unit #1 @ Wausau, WI	C	2.64	WU1WAWI
Weston Unit #2 @ Wausau, WI	C	2.64	WU2WAWI
Weston Unit #3 @ Wausau, WI	C	2.62	WU3WAWI
Xcel Energy @ Blackdog, MN (previously Eagan, MN)	C	2.66	XCLBLMN
Xcel Energy @ Riverside, St. Paul, MN	C	2.67	XCLSPMN

FIGURE D.42 Sources of fly ash.

AAN American Association of Nurserymen
AAR Association of American Railroads
AASHTO American Association of State Highway and Transportation Officials
AITC American Institute of Timber Construction
AC Alternating Current
ACI American Concrete Institute
AGC Associated General Contractors of America, Inc.
AIA American Institute of Architects
AISC American Institute of Steel Construction
AISI American Iron and Steel Institute
AITC American Institute of Timber Construction
ANSI American National Standards Institute
ARA American Railway Association
AREA American Railway Engineering Association
ASCE American Society of Civil Engineers
ASLA American Society of Landscape Architects
ASME American Society of Mechanical Engineers
ASTM American Society of Testing and Materials
ATR Automatic Traffic Recorder
AWPA American Wood Preservers Association
AWS American Welding Society
AWG American Wire Gauge
AWWA American Water Works Association
CCTV Closed Circuit Television
CMP Communications Plenum Cable or Corrugated Metal Pipe
CMS Changeable Message Sign
COAX Radio Frequency Transmission Cable (Coaxial Cable)
CRSI Concrete Reinforcing Steel Institute
CV Compacted Volume
DBE Disadvantage Business Enterprise
EEO Equal Employment Opportunity
EV Excavated Volume
EVP Emergency Vehicle Pre-Emption
FAA Federal Aviation Administration
FHWA Federal Highway Administration, U.S. Department of Transportation
FSS Federal Specifications and Standards, General Services Administration
FTA Federal Transit Administration
GFI Ground Fault Interrupter
HH Handhole
IEEE Institute of Electrical and Electronics Engineers
IES Illuminating Engineers Society
ICEA Insulated Cable Engineers Association
IMC Intermediate Metal Conduit
ISO International Standards Organization
IPS Iron Pipe Size
ITC Information Transmission Capacity
ITE Institute of Transportation Engineers

JMF Job Mix Formula used in the Bituminous Specifications.
KVA Kilovolt Ampere
LV Loose Volume for Measurements, or Leveling Course for Bituminous
MGal 1000 Gallons
MN MUTCD Minnesota Manual on Uniform Traffic Control Devices
Mn/DOT Minnesota Department of Transportation
MN Statutes Minnesota Statutes
MPCA Minnesota Pollution Control Agency
NEC National Electrical Code
NEMA National Electrical Manufacturers Association
NMC Non-Metallic Conduit
No. When reference is to wire, it is the AWG gauge number.
NPDES National Pollutant Discharge Elimination System
OSHA Occupational Safety & Health Administration
(P) (Defined in 1103)
PCI Prestressed Concrete Institute
PIV Peak Invert Voltage
PLS Pure Live Seed
PVC Polyvinyl Chloride
QA Quality Assurance
QC Quality Control
RCS Ramp Control Signal
REA Rural Electrification Association
RF Radio Frequency
RHW Moisture and Heat Resistant or Cross Linked Synthetic Polymer
RMS Root Mean Square
RSC Rigid Steel Conduit
SAE Society of Automotive Engineers
SI International System of Units (The Modernized Metric System)
SPDT Single Pole Double Throw
SPST Single Pole Single Throw
SSPC Society for Protective Coatings
SV Stockpiled Volume
SWPPP Storm Water Pollution Prevention Plan
TH Trunk Highway
TMC Traffic Management Center
TMS Traffic Management System
TSM Traffic System Management
UL Underwriters Laboratories, Inc.
USD United States Department of Agriculture
UV Ultra Violet
VAC Volt Alternating Current (60 Hz)
VDC Volt Direct Current
XHHW Moisture and Heat Resistant Cross Linked Synthetic Polymer

FIGURE D.43 Common abbreviations.

Inch = 1/12 or 0.083 foot = 2.54 centimeters = 25.4 millimeters
Foot = 12 inches = 0.3048 meters = 30.48 centimeters
Yard = 36 inches = 3 feet = 0.9144 meters
Rod = 16.5 feet = 5.5 yards = 5.03 meters
Furlong = 22 yards
Mile = 1,700 yards = 5,380 feet = 1.61 kilometers = 8 furlongs = 80 chains

FIGURE D.44 Length conversions.

Metric Prefix and Magnitude	Metric Symbols
M mega (10^6)	A ampere (electric current)
k kilo (10^3)	cd candela (luminous intensity)
m milli (10^{-3})	F farad (electric capacitance)
µ micro (10^{-6})	g gram (mass)
n nano (10^{-9})	H henry (inductance)
p pico (10^{-12})	ha hectare (area)
	Hz hertz (frequency - cycles or impulses per second)
	J joule (energy)
	km/h kilometer per hour (velocity)
	km^2 square kilometer (area)
	L liter (volume)
	m/s meters per second (velocity)
	m meter (length)
	m^2 square meter (area)
	m^3 cubic meter (volume)
	m^3/s cubic meters per second (flow rate)
	N newton (force)
	N•m newton meter (torque)
	Pa pascal (pressure, stress)
	s second (time)
	S siemens (electrical conductance)
	t metric ton (mass)
	V volt (electric potential)
	W watt (power)
	Ω ohm (electric resistance)
	°C degree Celsius (temperature)

FIGURE D.45 Metric measurements.

Problems	Potential Result(s)	Possible Cause(s)	What to Do
Poured joint sealant does not adhere	Water or incompressible materials may enter joint, causing poor joint performance	Joint face is dirty; joint shape factor is incorrect; concrete is too green when sealed and therefore contains too much moisture	Check joint face for cleanliness and dryness; check joint shape factor; replace sealer
Poured joint sealant fails (it is not cohesive)	Water or incompressible materials may enter joint, causing poor joint performance	Poor sealant properties due to over or under heating	Reduce heat; apply proper heat; use insulated hoses; replace sealant
Preformed sealant is loose	Water or incompressible materials may enter joint, causing poor joint performance	Sealant is not sized properly; joint too wide; stretched sealant	Use properly sized sealant (check joint width); check sealant quality; review installation procedure

FIGURE D.46 Potential problems with sealants.

Problems	Potential Result(s)	Possible Cause(s)	What to Do
Cracks form before sawing	Random, irregular cracks	Sawing too late	Immediately begin skip sawing: jump ahead to saw every second or third joint
Cracks form during sawing, in front of the saw	If sawcutting continues near a crack, the sawed joint will not function correctly; this could lead to spalling and other performance problems	Sawing too late	Immediately begin skip sawing: jump ahead to saw every second or third joint
Sawcut ravels during sawing	Spalled joints	Sawing too early	Stop sawing and wait for more strength to develop in the concrete
Joint face ravels or spalls	Poor joint performance over the long term	Sawcutting performed too early; poor sawcutting operation; joint area not cured properly	Review and correct sawcutting operations; review joint face curing process

FIGURE D.47 Sawing problems.

B-6 Construction Grout	Bonsal	*http://www.bonsalamerican.com*
F-77 Construction Grout	Bonsal	*http://www.bonsalamerican.com*
Burke Non-Ferrous, Nonshrink Grout	Burke	*http://www.burke.com*
Pro Grout 90	CGM Incorporated	*http://www.cgmbuildingproducts.com*
Pro Grout 100	CGM Incorporated	*http://www.cgmbuildingproducts.com*
Set Grout	Chemrex Inc.	*http://www.buildingsystems.basf.com*
100 Nonshrink, Nonmetallic Grout	Conspec	*http://www.conspecmkt.com*
Enduro 50	Conspec	*http://www.conspecmkt.com*
Sure-Grip High Performance Grout	Dayton-Superior	*http://www.daytonsuperior.com*
Por Rok	Hallemite	*http://www.arcat.com/arcatcos/cos32/arc32928.htm*
CRYSTEX	L & M Construction Chemicals	*http://www.lmcc.com*
Duragrout	L & M Construction Chemicals	*http://www.lmcc.com*
Sauereisen F 100	Sauereisen Cements	*http://www.sauereisen.com/c_prod/c_specialty.asp*
Superb Grout 611	Specco Industries Inc.	*http://www.specco.com*
Multipurpose Grout	Symons	*http://www.symons.com*
NC Grout	Tamms Industries	*http://www.euclidchemical.com/*
Uni Grout	Universal Form Clamp	*http://www.universalformclamp.com*
Uni Grout Super	Universal Form Clamp	*http://www.universalformclamp.com*
US Spec GP Grout	US Mix Corporation	*http://www.usmix.com*
US Spec MP Grout	US Mix Corporation	*http://www.usmix.com*
Certi-Grout # 1000	Vexcon Chemicals	*http://www.vexcon.com*
Pac-it	W. R. Meadows	*http://www.wrmeadows.com*
Sealtight 588	W. R. Meadows	*http://www.wrmeadows.com*

FIGURE D.48 Sources of nonshrink grout.

Metric	
1 centimeter =	10 millimeter
1 decimeter =	10 centimeter
1 meter =	10 decimeter
1 dekameter =	10 meter
1 hectometer =	10 dekameter
1 kilometer =	1000 m = 10 ha
1 Angstrom =	10^{-10} meter

English and Metric	
1 inch =	25.4 millimeter
1 feet =	12 inches
1 yard =	3 foot
1 rod/perch =	16.5 foot
1 chain =	22 yard
1 furlong =	40 rd = 220 yd
1 mile =	8 furlong
1 mile =	5280 feet
1 hand =	4 inches
1 mil =	0.001 inch

Nautical	
1 fathom =	6 feet
1 cable length =	720ft = 120 fathom
1 nautical mile =	1852 meter
1 league =	3 miles

FIGURE D.49 Conversion factors.

Cubic feet × 7.48 = Gallons

Cubic feet × 62.4 = Pounds of water

Gallons × 8.333 = Pounds of water

Gallons × 0.1337 = Cubic feet

Cubic inches × 1,728 = Cubic feet

Cubic yards × 27 = Cubic feet

Cubic feet × 27 = Cubic yards

FIGURE D.50 Volume conversion factors.

Cubic inch = 0.00058 cubic foot = 16.4 cubic centimeters

Cubic foot = 1,728 cubic inches = 0.037 cubic yard = 0.028 cubic meter

Cubic yard = 27 cubic feet = 0.765 cubic meters

FIGURE D.51 Volume conversions.

Metric: Mass Conversion Factors	
1 centigram =	10 milligram
1 decigram =	10 centigram
1 gram =	10 dg = 1000 mg
1 dekagram =	10 gram
1 hectogram =	10 dekagram
1 kilogram =	1000 gram
1 megagram =	1000 kilogram
1 megagram =	1 metric ton

Troy Mass Conversion Factors	
1 grain =	64.79891 milligram
1 pennyweight =	24 grains
1 ounce (oz t) =	20 pennyweight
1 pound (lb t)=	12 ounces

Avoirdupois Mass Conversion Factors	
1 grain =	64.79891 milligram
1 dram (dr) =	27.34375 grains
1 ounce (oz avdp) =	16 drams
1 pound (lb avdp) =	16 ounces
1 hundredweight =	100 pound
1 ton =	20 hundredweight
1 long hundredweight =	112 pound
1 long ton =	20 long hundredweight

Apothecaries Mass Conversion Factors	
1 grain =	64.79891 milligram
1 scruple(s) =	20 grains
1 dram (dr ap) =	3 scruple
1 ounce (oz ap) =	8 drams
1 pound (lb ap)=	12 ounces

Other Weight Conversion Factors	
1 carat (c) =	200 milligrams
1 gamma (y) =	1 microgram

FIGURE D.52 Weight conversions.

Weather	Characteristics	Possible Effect(s) and Problem(s)	What to Do
Hot and Dry	High air temperature (above 90°F or 32°C); low relative humidity; high wind speed; sunny	• High rate of water evaporation from mixture, especially with hot concrete (above 90°F) • Dry stockpiles • Rapid slump loss • Long-term strength loss due to added water • Rapid setting; less time for finishing	• If possible, do not pave in very hot, dry weather • Plan ahead (once problems are noticed, it may be too late to correct them) • Pave in the morning, evening, or night • Maintain stockpile moisture • Keep subbase and/or forms damp and cool • Keep equipment cool • Take extra care with curing; use additional curing compound
Cold	Low air temperature (below 50°F or 10°C)	• Low heat of hydration; very slow set • Frozen concrete mixture • Low strength gain • Increased concrete permeability	• If possible, do not mix or pave in very cold weather • Heat materials • Add accelerator to mixture • Increase Portland cement in mixture

FIGURE D.53 Working with concrete in hot and cold temperatures.

Glossary

Abrasion resistance: Ability of a surface to resist being worn away by rubbing and friction.

Acrylic resin: One of a group of thermoplastic resins formed by polymerizing the esters or amides of acrylic acid; used in concrete construction as a bonding agent or surface sealer.

Adhesives: The group of materials used to join or bond similar or dissimilar materials; for example, in concrete work, the epoxy resins.

Air-water jet: A high-velocity jet of air and water mixed at the nozzle; used in cleanup of surfaces of rock or concrete such as horizontal construction joints.

Alkali-aggregate reaction: Chemical reaction in mortar or concrete between alkalis (sodium and potassium) from Portland cement or other sources and certain constituents of some aggregates; under certain conditions, deleterious expansion of the concrete or mortar may result.

Alkali-carbonate rock reaction: The reaction between alkalis (sodium and potassium) from Portland cement and certain carbonate rocks, particularly calcitic dolomite and dolomitic limestones, present in some aggregates; the products of the reaction may cause abnormal expansion and cracking of concrete in service.

Alkali reactivity (of aggregate): Susceptibility of aggregate to alkali-aggregate reaction.

Alkali-silica reaction: The reaction between the alkalis (sodium and potassium) in Portland cement and certain siliceous rocks or minerals, such as opaline chert and acidic volcanic glass, present in some aggregates; the products of the reaction may cause abnormal expansion and cracking of concrete in service.

Autogenous healing: A natural process of closing and filling of cracks in concrete or mortar when kept damp.

Bacterial corrosion: The destruction of a material by chemical processes brought about by the activity of certain bacteria that may produce substances such as hydrogen sulfide, ammonia, and sulfuric acid.

Blistering: The irregular raising of a thin layer at the surface of placed mortar or concrete during or soon after completion of the finishing operation, or in the case of pipe after spinning; also bulging of the finish plaster coat as it separates and draws away from the base coat.

Bug holes: Small regular or irregular cavities, usually not exceeding 15 mm in diameter, resulting from entrapment of air bubbles in the surface of formed concrete during placement and compaction.

Butyl stearate: A colorless oleaginous, practically odorless material ($C17H35COOC4H9$) used as an admixture for concrete to provide dampproofing.

Cavitation damage: Pitting of concrete caused by implosion: the collapse of vapor bubbles in flowing water that form in areas of low pressure and collapse as they enter areas of higher pressure.

Chalking: Formation of a loose powder resulting from the disintegration of the surface of concrete or an applied coating such as cement paint.

Checking: Development of shallow cracks at closely spaced, but irregular, intervals on the surface of plaster, cement paste, mortar, or concrete.

Cold-joint lines: Visible lines on the surface of formed concrete indicating the presence of joints where one layer of concrete had hardened before subsequent concrete was placed.

Concrete, preplaced-aggregate: Concrete produced by placing coarse aggregate in a form and later injecting a Portland-cement-sand grout, usually with admixtures, to fill the voids.

Corrosion: Destruction of metal by chemical, electrochemical, or electrolytic reaction with its environment.

Cracks, active*: The cracks for which the mechanism causing the cracking is still at work. Any crack that is still moving.

Cracks, dormant*: Those cracks not currently moving or which the movement is of such magnitude that the repair material will not be affected.

Craze cracks: Fine, random cracks or fissures in a surface of plaster, cement paste, or mortar.

Crazing: The development of craze cracks; the pattern of craze cracks existing in a surface. (See also *Checking*.)

Dampproofing: Treatment of concrete or mortar to retard the passage or absorption of water or water vapor, either by application of a suitable coating to exposed surfaces or by use of a suitable admixture, treated cement, or preformed films such as polyethylene sheets under slabs on grade. (See also *Vapor Barrier*.)

D-cracking: A series of cracks in concrete near and roughly parallel to joints, edges, and structural cracks.

Delamination: A separation along a plane parallel to a surface as in the separation of a coating from a substrate or the layers of a coating from each other, or in the case of a concrete slab, a horizontal splitting, cracking, or separation of a slab in a plane roughly parallel to, and generally near, the upper surface; found most frequently in bridge decks and caused by the corrosion of reinforcing steel or freezing and thawing; similar to spalling, scaling, or peeling except that delamination affects large areas and can often be detected only by tapping.

Deterioration: Decomposition of material during testing or exposure to service. (See also *Disintegration*.)

Diagonal crack: In a flexural member, an inclines crack caused by shear stress, usually at about 45 degrees to the neutral axis of a concrete member; a crack in a slab, not parallel to the lateral or longitudinal directions.

Discoloration: Departure of color from that which is normal or desired.

Disintegration: Reduction into small fragments and subsequently into particles.

Dry-mix shotcrete: Shotcrete in which most of the mixing water is added at the nozzle.

Drypacking: Placing a zero slump concrete, mortar, or grout by ramming it into a confined space.

Durability: The ability of concrete to resist weathering action, chemical attack, abrasion, and other conditions of service.

Dusting: The development of a powdered material at the surface of a hardened concrete.

Efflorescence: A deposit of salts, usually white, formed on a surface, the substance having emerged in solution from within concrete or masonry and subsequently having been precipitated by evaporation.

Epoxy concrete: A mixture of epoxy resin, catalyst, and fine aggregate. (See also *Epoxy resin*.)

Epoxy resin: A class of organic chemical bonding systems used in the preparation of special coatings or adhesives for concrete or as binders in epoxy resin mortars and concretes.

Erosion: Progressive disintegration of a solid by the abrasive of cavitation action of gases, fluids, or solids in motion. (See also *Abrasion resistance* and *Cavitation damage*.)

Evaluation*: Determining the condition, degree of damage or deterioration, or serviceability and, when appropriate, indicating the need for repair, maintenance, or rehabilitation. (See also *Repair, Maintenance, Rehabilitation*.)

Exfoliation: Disintegration occurring by peeling off in successive layers; swelling up and opening into leaves or plates like a partly opened book.

Exudation: A liquid or viscous gel-like material discharge through a pore, crack, or opening in the surface of concrete.

Feather edge: Edge of a concrete or mortar patch or topping that is beveled at an acute angle.

Groove joint: A joint created by forming a groove in the surface of a pavement, floor slab, or wall to control random cracking.

Hairline cracks: Cracks in an exposed concrete surface having widths so small as to be barely perceptible.

Honeycomb: Voids left in concrete due to failure of the mortar to effectively fill the spaces among coarse aggregate materials.

Incrustation: A crust or coating, generally hard, formed on the surface of concrete or masonry construction or on aggregate particles.

Joint filler: Compressible material used to fill a joint to prevent to infiltration of debris and to provide support for sealants.

Joint sealant: Compressible material used to exclude water and solid foreign material from joints.

Laitance: A layer of weak and nondurable material containing cement and fines from aggregates, brought by bleeding water to the top of overwet concrete, the amount of which is generally increased by overworking or overmanipulating concrete at the surface by improper finishing or by job traffic.

Latex: A water emulsion of a high molecular-weight polymer used especially in coatings, adhesives, and leveling and patching compounds.

Maintenance*: Taking periodic actions that will prevent or delay damage or deterioration or both. (See also *Repair*.)

Map cracking: See *Crazing*.

Microcracks: Microscopic cracks within concrete.

Monomer: An organic liquid of relatively low molecular weight that creates a solid polymer by reacting with itself or other compounds of low molecular weight or with both.

Overlay: A layer of concrete or mortar, seldom thinner than 25 mm (1 in.), placed on and usually bonded onto the worn or cracked surface of a concrete slab to restore or improve the function of the previous surface.

Pattern cracking: Intersecting cracks that extend below the surface of hardened concrete; caused by shrinkage of the drying surface, which is restrained by concrete at a greater depth where little or no shrinkage occurs; vary in width and depth from fine and barely visible to open and well defined.

Peeling: A process in which thin flakes of mortar are broken away from a concrete surface, such as by deterioration or by adherence of surface mortar to forms as forms are removed.

Pitting: Development of relatively small cavities in a surface caused by a phenomenom such as corrosion or cavitation, or in concrete localized disintegration such as popout.

Plastic cracking: Cracking that occurs in the surface of fresh concrete soon after it is placed and while it is still plastic.

Polyester: One of a large group of synthetic resins, mainly produced by a reaction of dibasic acids with dihydroxyl alcohols, commonly prepared for application by mixing with a vinyl-group monomer and free-radical catalyst at ambient temperatures and used as bonders for resin mortars and concretes, fiber laminates (mainly glass), adhesives, and the like. (See also *Polymer concrete*.)

Polyethylene: A thermoplastic high-molecular-weight organic compound used in formulating protective coatings; in sheet form, used as a protective cover for concrete surfaces during the curing period, or to provide a temporary enclosure for construction operations.

Polymer: The product of polymerization; more commonly, a rubber or resin consisting of large molecules formed by polymerization.

Polymer-cement concrete: A mixture of water, hydraulic cement, aggregate, and a monomer or polymer polymerized in place when a monomer is used.

Polymer concrete: Concrete in which an organic polymer serves as the binder; also known as resin concrete; sometimes erroneously employed to designated hydraulic-cement mortars or concretes in which part or all of the mixing water is replaced by an aqueous dispersion of thermoplastic copolymer.

Polymerization: The reaction in which two or more molecules of the same substance combine to form a compound containing the same elements in the same proportions, but of higher molecular weight, from which the original substance can be generated, in some cases only with extreme difficulty.

Polystyrene resin: Synthetic resins varying in color from colorless to yellow formed by the polymerization of styrene, or heated, with or without catalysts, that may be used in paints for concrete or for making sculpted molds or as insulation.

Polysulfide coating: A protective coating system prepared by polymerizing a chlorinated alkypolyether with an inorganic polysulfide.

Polyurethane: Reaction product of an isocyanate with any of a wide variety of other compounds containing an active hydrogen group; used to formulate tough, abrasion-resistant coatings.

Polyvinyl acetate: Colorless, permanently thermoplastic resin, usually supplied as an emulsion or water-dispersible powder characterized by flexibility, stability toward light, transparency to ultraviolet rays, high dielectric strength, toughness, and hardness; the higher the degree of polymerization, the higher softening temperature; may be used in paints for concrete.

Polyvinyl chloride: A synthetic resin prepared by the polymerization of vinyle chloride; used in the manufacture of nonmetallic water-stops for concrete.

Popout: The breaking away of small portions of concrete surface due to internal pressure, which leaves a shallow, typically conical, depression.

Pot life: Time interval after preparation during which a liquid or plastic mixture is usable.

Reactive aggregate: Aggregate containing substances capable of reacting chemically with products of solution or hydration of the Portland cement in concrete or mortar under ordinary conditions of exposure, resulting in come cases in harmful expansion, cracking, or staining.

Rebound hammer: An apparatus that provides a rapid indication of the mechanical properties of concrete based on the distance of rebound of a spring-driven missile.

Rehabilitation: The process of repairing or modifying a structure to a desired useful condition.

Repair: Replace or correct deteriorated, damaged, or faulty materials components, or elements of a structure.

Resin: A natural or synthetic, solid or semisolid organic material of indefinite and often high molecular weight having a tendency to flow under stress that usually has a softening or melting range and usually fractures conchoidally.

Resin mortar (or concrete): See *Polymer concrete.*

Restraint (of concrete): Restriction of free movement of fresh or hardened concrete following completion of placement in framework or molds within an otherwise confined space; restraint can be internal or external and may act in one or more directions.

Rock pocket: A porous, mortar-deficient portion of hardened concrete consisting primarily of coarse aggregate and open voids, caused by leakage of mortar from form, separate (segregation) during placement, or insufficient consolidation. (See also *Honeycombing.*)

Sandblasting: A system of cutting or abrading a surface such as concrete by a stream of sand ejected from a nozzle at high speed by compressed air; often used for cleanup of horizontal construction joints or for exposure of aggregate in architectural concrete.

Sand streak: A streak of exposed fine aggregate in the surface of formed concrete that is caused by bleeding.

Scaling: Local flaking or peeling away of the near-surface portion of hardened concrete or mortar; also of a layer from metal. (See also *Peeling* and *Spalling*.)

Shotcrete: Mortar or concrete pneumatically projected at high velocity onto a surface; also known as air-blown mortar; pneumatically applied mortar or concrete, sprayed mortar, and gunned concrete. (See also *Dry-mix shotcrete* and *Wet-mix shotcrete*.)

Shrinkage: Volume decrease caused by drying and chemical changes; a function of time but not temperature or stress caused by external load.

Shrinkage crack: Crack due to restraint of shrinkage.

Shrinkage cracking: Cracking of a structure or member from failure in tension caused by external or internal restraints as reduction in moisture content develops or as carbonation occurs, or both.

Spall: A fragment, usually in the shape of a flake, detached from a larger mass by a blow, section of weather, pressure, or expansion within the larger mass; a small spall involves a roughly circular depression not greater than 20 mm in depth nor 150 mm in any dimension; a large spall may be roughly circular or oval or, in some cases, elongated more than 20 mm in depth and 150 mm in greatest dimension.

Spalling: The development of spalls.

Sulfate attack: Chemical or physical reaction, or both, between sulfates, usually in soil or in ground water and concrete or mortar, primarily with calcium aluminate hydrates in the cement-paste matrix, often causing deterioration.

Sulfate resistance: Ability of concrete or mortar to withstand sulfate attack. (See also *Sulfate attack*.)

Swiss hammer: See *Rebound hammer*.

Temperature cracking: Cracking as a result of tensile failure caused by temperature drop in members subjected to internal restraints.

Thermal shock: The subjection of newly hardened concrete to a rapid change in temperature that may be expected to have a potentially deleterious effect.

Thermoplastic: Becoming soft when heated and hard when cooled.

Thermosetting: Becoming rigid by chemical reaction and not meltable.

Transverse cracks: Cracks that develop at right angles to the long direction of a member.

Tremie: A pipe or tube through which concrete is deposited underwater, having at its upper end a hopper for filling and bail for moving the assemblage.

Tremie concrete: Subaqueous concrete placed by means of a tremie.

Tremie seal: The depth to which the discharge end of the tremie pipe is kept embedded in the fresh concrete placed in a cofferdam for the purpose of preventing the intrusion of water when the cofferdam is dewatered.

Vapor barrier: A membrane placed under concrete floor slabs that are placed on grade and intended to retard transmission of water vapor.

Waterstop: A thin sheet of metal, rubber, plastic, or other material inserted across a joint to obstruct seepage of water through a joint.

Water void: Void along the underside of an aggregate particle or reinforcing steel that formed during the bleeding period and initially filled with bleed water.

Weathering: Changes in color, texture, strength, chemical composition, or other properties of a natural or artificial material caused by the action of weather.

Wet-mixing shotcrete: Shotcrete in which ingredients, including mixing water, are mixed before introduction into the delivery hose; accelerator if used, is normally added at the nozzle

ABBREVIATIONS

ACI	American Concrete Institute
ASTM	American Society for Testing and Materials
AWA	antiwashout mixture
CDMS	Continuous Deformation Monitoring System
CE	Corps of Engineers
CERC	Coastal Engineering Research Center
CEWES-SC	U.S. Army Engineer Waterways Experiment Station, Structures Laboratory, Concrete Technology Division
CMU	concrete masonry units
CRD	Concrete Research Division, Handbook for Concrete and Cement
EM	Engineer Manual
EP	Engineer Pamplet
ER	Engineer Regulation
FHWA	Federal Highway Administration
GPS	Global Positioning System
HAC	high alumina cement
HMWM	high molecular weight methacrylate
HQUSACE	Headquarters, U.S. Army Corps of Engineers
HRWRA	high-range water-reducing admixture
ICOLD	International Commission on Large Dams
MPC	magnesium phosphate cement
MSA	maximum size aggregate
MSDS	Manufacturer's Safety Data Sheet
NACE	National Association of Corrosion Engineers
NCHRP	National Cooperative Highway Research Program
NDT	nondestructive testing
OCE	Office, Chief of Engineers
PC	polymer concrete
PCA	Portland Cement Association
PIC	polymer-inpregnated concrete
PMF	Probable Maximum Flood
PVC	polyvinyl chloride
R-values	rebound readings

RCC roller-compacted concrete
REMR Repair, Evaluation, Maintenance and Rehabilitation Research Program
ROV remotely operated vehicle
TM technical manual
TOA time of arrival
UPE ultrasonic pulse-echo
UV ultraviolet
w/c water–cement ratio
WES Waterways Experiment Station
WRA water-reducing admixture

Index

Printed in the United States
By Bookmasters